"This notable anthology crosses disciplinary, theoretical, and racial boundaries to showcase the importance of diverse garden types, as well as the collaborative efforts to create them, especially those of women. As underscored by the authors, the significance of women in shaping garden spaces of all kinds cannot be overstated."

Annette Giesecke, *Centre for Science in Society, Victoria University of Wellington | Te Herenga Waka, New Zealand*

"The contributors to this volume view gardens as inherently collective ventures that help us understand our place in the world. Their case studies reference multiple media to document women's contribution to gardening from antiquity to the present day. The collection will please and inform readers while advancing the feminist cause."

Stephanie Ross, *University of Missouri-St. Louis, USA*

W0234855

Women and the Collaborative Art of Gardens

Women and the Collaborative Art of Gardens explores the garden and its agency in the history of the built and natural environments, as evidenced in landscape architecture, literature, art, archaeology, history, photography, and film.

Throughout the book, each chapter centers the act of collaboration, from garden clubs of the early twentieth century as powerful models of women's leadership, to the more intimate partnerships between family members, to the delicate relationship between artist and subject. Women emerge in every chapter, whether as gardeners, designers, owners, writers, illustrators, photographers, filmmakers, or subjects, but the contributors to this dynamic collection unseat common assumptions about the role of women in gardens to make manifest the significant ways in which women write themselves into the accounts of garden design, practice, and history. The book reveals the power of gardens to shape human existence, even as humans shape gardens and their representations in a variety of media, including brilliantly illuminated manuscripts, intricately carved architectural spaces, wall paintings, black and white photographs, and wood cuts. Ultimately, the volume reveals that gardens are best apprehended when understood as products of collaboration.

The book will be of interest to scholars and students of gardens and culture, ancient Rome, art history, British literature, medieval France, film studies, women's studies, photography, African American studies, and landscape architecture.

Victoria E. Pagán is Professor of Classics at the University of Florida, USA.

Judith W. Page is Professor of English and Distinguished Teaching Scholar Emerita at the University of Florida, USA.

Routledge Environmental Literature, Culture and Media
Series Editor: Thomas Bristow

The urgency of the next great extinction impels us to evaluate environmental crises as sociogenic. Critiques of culture have a lot to contribute to the endeavour to remedy crises of culture, drawing from scientific knowledge but adding to it arguments about agency, community, language, technology and artistic expression. This series aims to bring to consciousness potentialities that have emerged within a distinct historical situation and to underscore our actions as emergent within a complex dialectic among the living world.

It is our understanding that studies in literature, culture and media can add depth and sensitivity to the way we frame crises; clarifying how culture is pervasive and integral to human and non-human lives as it is the medium of lived experience. We seek exciting studies of more-than-human entanglements and impersonal ontological infrastructures, slow and public media, and the structuring of interpretation. We seek interdisciplinary frameworks for considering solutions to crises, addressing ambiguous and protracted states such as solastalgia, anthropocene anxiety, and climate grief and denialism. We seek scholars who are thinking through decolonization and epistemic justice for our environmental futures. We seek sensitivity to iterability, exchange and interpretation as wrought, performative acts.

Routledge Environmental Literature, Culture and Media provides accessible material to broad audiences, including academic monographs and anthologies, fictocriticism and studies of creative practices. We invite you to contribute to innovative scholarship and interdisciplinary inquiries into the interactive production of meaning sensitive to the affective circuits we move through as experiencing beings.

Literary Feminist Ecologies of American and Caribbean Expansionism
Errand into the Wilderness
Christine M. Battista and Melissa R. Sande

Women and the Collaborative Art of Gardens
From Antiquity to the Present
Edited by Victoria E. Pagán and Judith W. Page

For more information about this series, please visit: www.routledge.com/
Routledge-Environmental-Literature-Culture-and-Media/book-series/RELCM

Women and the Collaborative Art of Gardens

From Antiquity to the Present

Edited by
Victoria E. Pagán and Judith W. Page

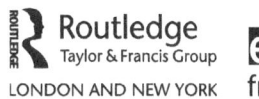

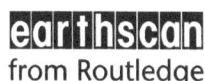

LONDON AND NEW YORK

First published 2024
by Routledge
4 Park Square, Milton Park, Abingdon, Oxon OX14 4RN

and by Routledge
605 Third Avenue, New York, NY 10158

Routledge is an imprint of the Taylor & Francis Group, an informa business

British Library Cataloguing-in-Publication Data
A catalogue record for this book is available from the British Library

Library of Congress Cataloging-in-Publication Data
Names: Pagán, Victoria Emma, editor. | Page, Judith W., editor.
Title: Women and the collaborative art of gardens: from antiquity to the present / edited by Victoria E. Pagán and Judith W. Page.
Description: Abingdon, Oxon; New York, NY: Routledge, 2024. |
Includes bibliographical references and index.
Identifiers: LCCN 2023026234 (print) | LCCN 2023026235 (ebook) |
ISBN 9781032464077 (hardback) | ISBN 9781032464091 (paperback) |
ISBN 9781003381549 (ebook)
Subjects: LCSH: Gardens—History. | Gardening—Social aspects. |
Gardens in literature. | Gardens in art. | Female friendship. | Artistic collaboration.
Classification: LCC SB451 .W65 2024 (print) | LCC SB451 (ebook) |
DDC 635.09—dc23/eng/20230825
LC record available at https://lccn.loc.gov/2023026234
LC ebook record available at https://lccn.loc.gov/2023026235

ISBN: 978-1-032-46407-7 (hbk)
ISBN: 978-1-032-46409-1 (pbk)
ISBN: 978-1-003-38154-9 (ebk)

DOI: 10.4324/9781003381549

Typeset in Times New Roman
by codeMantra

To Andy and Bill

Contents

List of Contributors   *xi*
Acknowledgments   *xiii*

Introduction   1
VICTORIA E. PAGÁN

1   **Garden Design as Feminist Ground**   10
    THAÏSA WAY

2   **Pompeian Gardens and the Archaeological Imagination**   34
    BETTINA BERGMANN

3   **The Garden's Transformational Artifice in Valois France**   66
    ELIZABETH ROSS

4   **Garden Theory, Gardening Practice: William and Dorothy
    Wordsworth**   91
    JUDITH W. PAGE

5   **Places for the Spirit, Photographs of Traditional African
    American Gardens**   112
    VAUGHN SILLS

6   **On the Diagonal, through the Window: Marie Menken's
    Glimpse of the Garden, 1957 and Rosalind Nashashibi's
    Vivian's Garden, 2017**   129
    MAUREEN TURIM

7   **Virginia Woolf and Vanessa Bell at Kew Gardens**   145
ELISE L. SMITH

Epilogue: What If We Start with the Garden?   171
JUDITH W. PAGE

Index   *177*

Contributors

Bettina Bergmann is Professor Emerita of Art History at Mount Holyoke College, USA. She has published numerous essays on the décor of Roman villas and houses, on ancient landscape and gardens, and on representations of women. Recently she has been working on the modern discovery and reception of Pompeii.

Victoria E. Pagán is Professor of Classics at the University of Florida, USA. She is the author of *Rome and the Literature of Gardens* (London, 2006). With Judith W. Page and Brigitte Weltman-Aron she co-edited *Disciples of Flora: Gardens in History and Culture* (Newcastle upon Tyne, 2015).

Judith W. Page is Professor of English and Distinguished Teaching Scholar Emerita at the University of Florida, USA. She is the author of articles and books on Romanticism and, most recently, of two books on women and gardens, both published by Cambridge University Press and co-authored with Elise L. Smith.

Elizabeth Ross is Associate Professor of Art History at the University of Florida, USA. She has written *Picturing Experience in the Early Printed Book*: *Breydenbach's Peregrinatio from Venice to Jerusalem* (University Park, 2014). She is currently the Director of the School of Art and Art History.

Vaughn Sills is an artist and Associate Professor Emerita, Art and Music Department, Simmons University, USA. Her photographs are in collections including the Smithsonian American Art Museum and Harvard Art Museum. Her publications include *Places for the Spirit, Traditional African American Gardens* (Trinity University Press, 2010) and *One Family* (UGA Press, 2001).

Elise L. Smith is Professor Emerita of Art History, Millsaps College, USA. Among her publications is a monograph on Evelyn De Morgan. With Judith Page she has co-authored two books on British women in the garden (Cambridge, 2011 and 2021) and a completed manuscript on gardening women in the American South.

Maureen Turim is Professor Emerita of Film and Media Studies in the Department of English at the University of Florida, USA. She is the author of three books and over a hundred essays in books and journals, writing often on feminist theory, spatial representation, and avant-garde and artists' films.

Thaïsa Way FASLA, FAAR, is the Director of Garden and Landscape Studies at Dumbarton Oaks Research Library and Collection, a Harvard University research institution in Washington DC and Professor Emerita, University of Washington, USA. Her most recent book is co-edited with Eric Avila, *Segregation and Resistance in American Landscapes* (2023).

Acknowledgments

This collection has been several years in the making and along the way we have benefited from collaborating with many scholars, colleagues, and institutions. Like the gardens that we and our contributors have studied, this book has grown from many seeds and has been nurtured by many hands. The anonymous reviewers for Routledge have helped us shape the volume into a coherent whole, and our contributors have shown their commitment to the project and their flexibility as it has developed.

For their financial support of this project, we would like to thank the Gladys Krieble Delmas Foundation. We are also grateful for the financial support from the following units at the University of Florida: the Center for the Humanities and Public Sphere; the Harn Eminent Scholar Chair in Art History; the Office of Research; the College of Liberal Arts and Sciences; the Department of Classics; the Department of English; the Center for Gender, Sexualities, and Women's Studies Research; and the Center for Greek Studies.

For their assistance, we thank Maura Gleeson, Erika Railey, Dulce Roman, Eric Segal, Donna Tuckey, and Rachel Wayne.

We thank the editorial staff at Routledge for their diligence and support, especially Grace Harrison, who was our editor in the early stages of the project, Rosie Anderson, who continued as editor, and Matthew Shobbrook, who has always answered our questions with alacrity and professionalism. Tom Bristow, the series editor, has provided encouragement and expertise.

The contributors wish to thank the staff of the Millsaps College Library and the British Library; the Polaroid Corporation, the Polaroid Foundation, and Simmons College; Carlyn Ferrari, Seattle University and Andrea Roberts, University of Virginia; Riccardo Helg and Gülru Çackmak; and Alivia Hunter.

Finally, William H. Page and Andrew Wolpert have served as sounding boards for so many ideas and provided essential moral support during this process.

Introduction

Victoria E. Pagán

Writing against the backdrop of decades of brutal civil war, the Roman poet Vergil composed the *Georgics* during the age of Augustus, Rome's first emperor. The poem provides solid evidence for how gardens are implicated in political identity and power. The core of the poem is a retreat from violence and its ugliness, and its perfection is derived in part from its distilled refusal to represent history, its grandeurs and nightmares. Yet it is probably not political office, or dictatorship, or even civil war that frightens the poet-gardener of the *Georgics* but the more immediate, unnamed, yet very real potential to be driven off one's land, to lose one's home, or to be sent into exile, even though the poet Vergil probably never envisaged or experienced such fates himself.

In the fourth and final book of the *Georgics* (4.116–148), Vergil provides a relatively short and deliberately truncated description of a garden belonging to a solitary old man who tended a plot "not rich enough for plowing, unsuitable for flocks, unfavorable for grapevines" (4.128–129). This problematic piece of land, incapable of producing economic benefit, becomes the ideal spot for a garden. The old man planted herbs, lilies, and vervain, and he harvested roses in the spring, apples in the autumn; even in winter he could find hyacinths blooming. His trees bore fruit such as pears and plums, and the plane tree provided shade for drinking (4.130–146). The scene is reminiscent of a Golden Age, but with one added component: a human being. This old Corycian, who tends his garden under the walls of Tarentum on the banks of the Galaesus River in southern Italy, has been identified as a historical person known to Vergil, a politician, a philosopher, or even a pirate.[1] He has also been interpreted as a product of pure imagination, a model of simplicity, or even as Orpheus, the mythological poet who charmed his way to the Underworld, and whose tragic story of the loss of his dear wife Eurydice brings the *Georgics* to a close. Regardless of the identity of the old man, the description obfuscates the tremendous amount of work it takes to plant and tend this garden, just as it obfuscates the know-how required—the purported content and intent of the entire poem. Instead, the description of this garden crystalizes, in the words of W. Ralph Johnson, "the Epicurean sage in his proper landscape, caught in the exact moment of the strange inactive action that best characterizes him."[2] The old man is independent, serene, lacking nothing, needing only what he has and having only what he needs: he works—and works relentlessly, it would seem—alone.

DOI: 10.4324/9781003381549-1

In short, as a central figure in the *Georgics,* he is a colossal fiction, and like most of the *Georgics*, he is impossibly imaginary. Our book altogether defies this venerable paradigm of the self-contained. Over and against the solitary old man who works alone, whose garden is isolated and idealized, we propose the collaboration of women, whose gardens are shared and realized and therefore capable of conveying political, social, and cultural meanings hitherto unexplored.

As the contributions in Sue Edney and Tess Somervell's recent addition to the Routledge Environmental Literature, Culture and Media Series have argued, the Georgic genre continues to serve as a way of thinking about humans' relationships with the non-human world, a genre that explores environmental ethics and its liabilities.[3] Indeed, literature is one of the most prominent media in which gardens are expressed. Drawing on Homer, Boccaccio, Italo Calvino, Andrew Marvell, Hannah Arendt, Rihaku, Karel Čapek, Coa Zuequin, Ariosto, and Michel Tournier, the Qur'an, and the Bible, Robert Pogue Harrison brings together religion, politics, education, therapy, and environmental responsibility.[4] He adroitly mines these diverse literary sources to uncover the garden as a metaphor for the human condition, and in so doing he legitimates the metaphysics of gardens. However, Harrison does not engage directly with visual arts, and observations about gender and race are subordinate to the thesis of the human condition.

Our volume is an extension of this work, but with a deliberately more expansive remit that moves the discussion beyond literature and the general human condition to other art forms and more intersectional concerns. Scholars of ancient Rome, art history, British literature, medieval France, film studies, women's studies, and landscape architecture are put in dialogue with a practicing photographer, but the overarching framework of collaboration demands that no one discipline or medium is privileged above the other. By crossing historical, generic, and theoretical boundaries and by putting multiple disciplines and media in conversation with each other, we open new vistas for understanding the nature and power of gardens to shape human existence, even as humans shape gardens and their representations in a variety of media. Scholars have long recognized that temporality, representation, political power, and gender are essential elements of gardens.[5] Our aim is to expose a productive tension between documentation and interpretation at work in the textual and visual mediations of a garden's life. In talking about gardens as well as making gardens, actual gardens and their representations in a wide variety of arts, we present an approach that affords a glimpse of their otherwise intangible essence.

Gardens are inherently collective ventures. As Elizabeth Lawrence recognizes, "No one can garden alone."[6] Garden clubs, folk gardens, community gardens, horticultural societies, and botanical garden memberships are just some of the formal mechanisms for the collaborative art and the starting point for historical public contexts. Gardens are also created and maintained by means of private entities, family relationships, and even friendships. Further still, gardens are the products of a collaboration between human agency and the natural world. Yet to what extent are gardens emblematic of the environment more generally, and what can collaboration contribute to discussions about the environmental crisis? The studies gathered here add depth and sensitivity to the way the environment is framed to apprehend structures of interpretation. Just as gardens are microcosms of the natural world, so this book is part of a series that addresses the urgency of the present environmental crisis. To exclude gardens from that critical conversation is to leave the ground—the plots of gardens large and small, and their potent ideologies—in the hands of government, corporate, and large-scale enterprises. Gardens connect human endeavors to the earth. Gardens conserve and restore the environment. The act of gardening returns agency to individuals and communities. Metaphysically, gardens draw together diverse discourses and philosophical assumptions that are central to how we understand our place in the world, and the extent to which it is ethical for us to shape that place. In keeping with the themes of the series, then, this book aims to provide an overview of the different ways collaboration underpins these human relationships with the environment, from first-century Rome to the twenty-first century, from Europe to Guatemala to the American South, through focused case studies on individual texts and artifacts. The book is the product of scholarly teamwork; however, the individual chapters themselves combine different methodologies, genres, media, and in the case of Chapter 2, even time periods. Gardens require analysis rooted in cooperation.

Collaboration is an edgy term, deliberately chosen. On the one hand, in certain historical contexts, whether voluntary or involuntary, it conveys a negative valence, akin to fraternization, collusion, even treason. For example, Vichy France of 1940–1944, officially independent, existed under German occupation; legitimacy was not unquestioned. To set aside historical political examples is to establish a spectrum, whose other end holds the promise of increased possibilities for human achievement. The cooperative efforts of the Wright brothers to develop the three-axis controls that made a fixed wing aircraft possible resulted in one of the most formative inventions of the twentieth century. We speak of partnerships in hendiadys (Ben and Jerry's ice cream; Simon and Garfunkel's music) that creates a willing suspension of the reality that the balance in any partnership, at some point, is going to be asymmetrical.

Collaboration can look very different depending on the participants and the terms of agreement. It involves the careful negotiation of power and an acceptance that one party may hold more power or influence than the other at any given time. Ideally the shift in power is balanced; however, the failure to observe this balance is a trope all too familiar. For example, in Greek mythology, the rule of Thebes is divided between the sons of Oedipus, but Eteocles refuses to share the annual rotation of the kingship of Thebes with his brother Polynices. However, by centering his play on Antigone and dramatizing her suffering due to the breakdown of the patriarchy and refusal to cooperate, Sophocles in effect genders collaboration, with recognizable implications even today. The teamwork of Watson and Crick laid the foundation for understanding the human genome; together with Wilkins they shared the Nobel Prize for Physiology or Medicine in 1962. Yet not one of these men acknowledged the crucial contribution of Rosalind Franklin, whose catalytic role in unravelling the structure of the DNA double helix was largely ignored until 1975.[7] Because of Franklin and her collaborators, scientists today can use DNA sequencing to track the ever-changing strains of COVID.

Gendered collaboration emerges full force in our volume. Elise Smith unpacks the work between Virginia Woolf and her sister Vanessa Bell in the creation of the published editions of the short story "Kew Gardens" in small hand-printed booklets illustrated with woodcuts, printed by the Hogarth Press. The effort of the sisters reveals conflicted views of the garden as claustrophobic and liberating. Maureen Turim analyzes the cooperation documented in the experimentally conceived film, *Vivian's Garden*, between artist Vivian Suter and nature itself, as the artist resigns her canvases to the elements and leaves them in the rain. Vaughn Sills' visual essay manifests the relationship born of trust between herself, the photographer, and her subjects; both are radically changed by the experiences of creating gardens and recreating the gardens in photographs. In this way, the traditional divide between gardens and their representations begins to erode, as the one is seen to influence the direction of the other. Women appear in every chapter, whether as denizens, gardeners, designers, owners, writers, illustrators, photographers, or filmmakers. The special relationship between women and gardens is self-evident; however, the chapters here make visible the variety of significant ways in which women write themselves into the accounts of garden design, practice, and history. Collaboration is a feminist issue insofar as it depends on acknowledging power imbalances and, more to the point, working toward a more just and inclusive world.

Feminist theorists connect care for the earth and the environment with intersectional categories of gender and race. Greta Gaard sketches the pivot from feminist ecocriticism to ecofeminist literary criticism that explores texts "to illuminate ways that Euro-American conceptions of 'nature' contain implicit formulations of gender, race, and indigeneity" and how these texts advance "alternative conceptions of nature." This critical approach has provided fresh analyses of women authors; recuperated earlier feminist works; foregrounded women's environmental justice literature; analyzed colonialism to argue for culture change; opened the way for queer and transgender ecocritical approaches. She concludes:

> Feminist environmental literature and ecocriticism affirm that a different world is possible: In the final decade before climate change outcomes become irreversible, the vision and goals of a feminist ecocriticism articulate a queer, multispecies, and antiracist feminist climate justice – a complete shift in economics, cultural values and views, governance, urban planning, energy, food systems, and, most of all, narratives. For the stories we tell about our origins and our dreams, about who counts as family and community, have everything to do with the eco-justice transformations we create.[8]

It is our view that because gardens express our origins and dreams, our family and community, they play a critical role in these eco-justice transformations. Although our collection does not address all these issues directly, we show how gardens introduce a broad range of political and cultural concerns. And so, although the content of this volume includes white upper-class British and European subjects, our book also puts, for example, the elite Roman gardens of ancient Pompeii in dialogue with gardens of the African American South. In her analysis of film, Maureen

Turim explores Indigenous issues in Guatemala; and Judith Page documents Dorothy Wordsworth's interest in the cottage garden, which was not necessarily elite and was more accessible to middle- and working-class society.

After reading the contributions in this volume, three conclusions will be inescapable. First, it will be impossible to conceive of gardens as endeavors of individuals; this book establishes gardens as a collaborative art. Second, while other books recognize the importance of gender and women in the study of gardens, in this volume, we center women as collaborators in the creative process, not least because it is the product of an interdisciplinary team of scholars working with a practicing artist. And third, the historical breadth, from antiquity to the present, together with the variety of media, including visual arts, literature, photography, and film, allows for the application of our conclusions beyond traditional white upper-class gardens of a specific time and specific geographical setting.

Driving the first chapter, "Garden Design as Feminist Ground," by Thaïsa Way, is a feminist impulse to bring to light the role of women in the development of the practice of landscape architecture and garden design as an art and a science in the twentieth century. While she places women at the center of her inquiry, Way simultaneously interrogates the intersections of race.[9] She begins by considering garden design as the art of modifying the natural world to create a garden room. In a manner reminiscent of Virginia Woolf's call for *A Room of One's Own* (1929), to design a garden does not merely position one as a professional but it is a radical act of creating a place of one's own that for contemporary women was considered not merely safe, but deeply moving and creative. Way subverts the idea of the professional as serving the public at large and suggests in addition that women garden designers may have been equally interested in fostering the well-being and creativity of their fellow women, gardeners, and garden enthusiasts.

In Chapter 2, "Pompeian Gardens and the Archaeological Imagination," Bettina Bergmann traces how the early excavations of the ancient gardens of Pompeii, the Roman city buried under volcanic ash and pumice in the eruption of Mount Vesuvius in 79 CE, inspired inventive recreations by the contemporary painters Lawrence Alma-Tadema (1836–1912) and Luigi Bazzani (1836–1927). Collaborating with archaeologists and using the material remains of excavated gardens, these male artists approximated historical accuracy even as they animated their scenarios with fictive activities of women among flowers that feminized the open spaces. Their quasi-photographic renderings offer illuminating precursors to modern digital reconstructions and immersive museum installations that transport viewers into an idyllic past. Fabrication engenders fabrication as the garden moves almost seamlessly from medium to medium, from the reality of 79 CE (known only to us by reconstruction) to the virtual reality of the twenty-first century.

The transformations traced by Bergmann are echoed in Chapter 3, "The Garden's Transformational Artifice in Valois France," by Elizabeth Ross, who examines the garden and visual representations of gardens in early Valois France (circa 1365–1420). Ross offers rich analyses of brilliantly illuminated manuscripts, intricately carved architectural spaces, wall paintings, and tapestries—all interior arts that are transformed by their representations of the gardens of outdoors. In these artifacts, the conceptual richness of the garden as a liminal site of transformation becomes a platform for expressing a conception of courtly life. The social code and the garden, the courtly and the courtyard, are one and the same: the prerogative of the socially and politically powerful elite—elite who are seen, in Ross's analysis, to collaborate in their use of imagery intended to solidify their standing. Underpinning the imagined space of the garden as an emblematic space of courtliness is the tacit cooperation between the lord and his peasants, a relationship easily transferred to God and his people. Garden imagery in late medieval art affords yet another glimpse of the asymmetry of collaboration.

After Ross's exploration of the many ways that gardens are represented in medieval French art, Judith Page's Chapter 4, "Garden Theory, Gardening Practice: William and Dorothy Wordsworth," excavates the letters of Wordsworth in which he outlines his garden theory and design plans and contemplates the relationship between nature and art, wild and cultivated. Furthermore, Page draws attention to the views of Wordsworth on beauty and symmetry, which he shared with his sister and some-time collaborator, Dorothy Wordsworth. Wordsworth anticipates some influential modern theories of what gardens mean and how they function aesthetically, including the notions of the afterlife of gardens as propounded by John Dixon Hunt.[10] And finally, the Wordsworths provide a framework for understanding the garden as not just a place of visual delight but, in the words of Page, "a place that embodies diurnal and seasonal change, a sense of the passage of time and the cycles of life. ... The garden thus provides a poignant reminder of our essential humanity."[11] The changes and reformulation of gardens after the designer's work is done, in alternative media that encompass multiple subjectivities, are central to each of the chapters in this volume that demonstrate that the power of gardens, affective and political, originates in the matrix of artifice and identity. Although gardens are above all enclosures, they are best comprehended when traditional disciplinary boundaries are broken.

In Chapter 5, we see the artist at work. "Places for the Spirit, Photographs of Traditional African American Gardens" is Vaughn Sills' personal narrative accompanied by a visual essay in photographs through which we can simultaneously perceive the artistic drive and its manifestation. One might object that Sills photographed yards, not gardens. However, the shadowy etymology of the English word "yard" is closely affiliated with Germanic root words that ultimately derive from *ghort*; yards, too, share the same lexical semantics of courts and enclosed spaces and descend perhaps from the same essential linguistic origin. Gender, race, and class intersect in this stimulating visual essay which chips away at the traditional divide between theory and practice. If Bergmann's chapter gives us quasi-photographic paintings by men who rendered inventive recreations of Pompeii—representations of male elite Roman culture by male elite European artists—then Sills' photographs, on the other hand, are representations of African American gardens, many created by women and recreated by a woman.

All these themes are carried into the final two chapters that treat material from the twentieth century. In Chapter 6, "On the Diagonal, through the Window: Marie Menken's *Glimpse of the Garden*, 1957 and Rosalind Nashashibi's *Vivian's*

Garden, 2017," Maureen Turim turns our attention to the themes of temporality, gender, and political power through her study of Marie Menken's film, *Glimpse of the Garden* (1957) and Rosalind Nashashibi's *Vivian's Garden* (2017). The two films, made sixty years apart, provide two very different paradigms for examining women's garden-inspired artmaking. Camera work and editing reconstitute the garden and make one work of art into another. *Vivian's Garden* documents émigré artists Elisabeth Wild (born 1922, Vienna) and her daughter Vivian Suter (born 1945, Buenos Aires) as they practice the arts of collage and painting. After Suter's paintings were damaged by mudslides during Hurricane Stan in 2005, she began to regard the effects of the hurricane not as damage but as part of the life of the paintings themselves. Since then, Vivian began purposely leaving her paintings in the open air, thus collaborating with nature in the process of making art. The film, *Vivian's Garden*, establishes a view of women artists in the garden that parallels— and departs from—Menken's montage of Dwight Ripley's garden. Both films present gardens as sites of refuge and danger. Nashashibi sets the film in Guatemala, where the mother and daughter lived in fear of rival drug gangs fighting in the area. Menken's audio-visual aesthetic is characterized by manic camera work and poorly recorded birdsong sampled from a stock source that plays eerily and unnaturally in a continuous loop. While the films mediate the experience of the individual gardens, they also allow us to move within the gardens. The gardens in these films—or should I say, the films of these gardens—insist that the viewer admit the menacing realities of our wholesale myths of emotional well-being. As Sills takes responsibility for her creations of African American gardens, so Turim's chapter shows how the filmmakers ask of their viewers a responsibility for all that their gardens have to offer as both refuge and risk.

While Turim treats two different films, in Chapter 7, "Virginia Woolf and Vanessa Bell at Kew Gardens," Elise Smith studies two editions of the same work of art, thereby shortening the interpretive gap. Virginia Woolf published two editions of her short story "Kew Gardens" (composed in 1917) in small hand-printed booklets illustrated with woodcuts by her sister Vanessa Bell. The story is told from the point of view of the creatures that reside in a small oval flower bed within the grand botanical garden. First appearing in 1919 with two illustrations, a new edition came out in 1927 with decorative borders on each page of text. The collaboration between the sisters reveals conflicted views of the garden. As Vivian Suter's garden was both refuge and danger, so Woolf's Kew Gardens is both claustrophobic and liberating. For some, "Kew Gardens" is an exercise in pure formal experimentation in the modernist mode, but for others the booklet has a political subtext, for clearly Woolf subverts the implicit imperial dominance of the grand botanical garden as an instrument of colonial expansion by showing its fragility. Woolf's "atmospheric story," as Smith calls it, merely alludes to specimens intended to assert imperial domination; more important in the story are the garden visitors' private conversations about war and the realization in the end that just beyond the garden is the city of London. To this, Vanessa Bell's woodcuts are for Smith, "best considered extrapolations rather than illustrations." The woodcuts do not represent the story; rather, somewhere between illustration and decoration, the woodcuts amplify the story. As the product of a physical and metaphysical collaboration, the booklet is far more than a mere representation or recreation of Kew Gardens in word or in image.

In the end, this book makes a singular demand of its reader, to withhold drawing synchronic momentary conclusions to perceive the complementary diachronic functions at work in the hermeneutics of the garden. Our approach deliberately militates against the very hierarchies of a "Table of Contents" that imposes a necessarily linear progression of thought. The reader will be challenged by the expansive and unique contexts of production collated here, ranging in time from the late Roman republic to the present. No doubt individual chapters will appeal to their respective experts. However, this broad disciplinary variety does more than merely showcase different approaches to the study of gardens; it demonstrates the strength of a model that is applicable across space and time. Collaboration is manifested in the representations of gardens in film, photography, painting, and many other arts. Of course, representation has been for decades a foundational tool for the study of the humanities.[12] Because a representation in any medium cannot capture every single aspect of the garden it attempts to record, some elements remain undefined, for which the artist substitutes continuity for the observer to infer. Thus, representations of gardens are so insistently fictional as to become paradoxically realistic: the represented garden is only what it looks like, a garden. The undefined inferences thus come to represent the process of representation. For gardens, this process is organic; although gardens persist and even have an afterlife, they are nonetheless constantly changing. Today's garden is a simulacrum of yesterday's; its uniqueness is fueled by the very norms of its existence. Yet the collective endeavors of the scholars in this book move the inquiry well beyond such a singularly theoretical approach that results in a lifeless discourse, in which interpretations fold in upon themselves in a series of self-referential, and ultimately hollow, metapoetics. Instead, we propose to correct and vivify this tendency by simultaneous attention to the practice of collaboration. Although the artificially imposed sequence of these chapters necessarily propels the reader, nevertheless, the cumulative investigation confronts contemporary assumptions of the unfolding of time and meaning, thus challenging narratives of the garden plot. Our collaboration reveals gardens and the discourses surrounding them as recognizable and unfamiliar, thus advancing a fresh approach and new pathways of knowledge.

Notes

1 Thomas, "The Old Man Revisited," 35.
2 Johnson, "A Secret Garden," 78.
3 Edney and Somervell, *Georgic Literature and the Environment*, 2, 3.
4 Harrison, *Gardens: An Essay on the Human Condition*.
5 For example, on temporality, Pugh, *Garden—Nature—Language* offers one of the earliest cultural interpretations of gardens, and his insights still resonate. On representation, Ross, *What Gardens Mean* explores the landscape gardens of eighteenth-century England to document the various relations between gardens and the art of painting. On political power, the essays in Giesecke and Jacobs *The Good Gardener* are excellent studies of power dynamics in gardens from a variety of historical periods. On gender,

Moore, *Sister Arts* explores the tradition of the erotic garden to create art with and for other women; more generally, Parker, "Unnatural History."

6 Lawrence, *The Little Bulbs*, 5.
7 Maddox, *Rosalind Franklin: The Dark Lady of DNA*.
8 Gaard, *"Nature, Gender, Sexuality,"* 265, 267–73, 274.
9 Drawing on the work of Crenshaw, *"Demarginalizing the Intersection."*
10 Hunt, *The Afterlife of Gardens*.
11 Page, in this volume, p. 96.
12 Wells, *The Dialectics of Representation*.

References

Crenshaw, Kimberlé. "Demarginalizing the Intersection of Race and Sex: A Black Feminist Critique of Antidiscrimination Doctrine, Feminist Theory, and Antiracist Politics [1989]," in K. Bartlett and R. Kennedy, eds., *Feminist Legal Theory*, London: Routledge, 1991: 57–80.

Edney, Sue and Tess Somervell. *Georgic Literature and the Environment: Working Land, Reworking Genre*, New York: Routledge, 2023.

Gaard, Greta. "Nature, Gender, Sexuality," in Peter Remien and Scott Slovic, eds., *Nature and Literary Studies*, Cambridge: Cambridge University Press, 2022: 261–79.

Giesecke, Annette and Naomi Jacobs. *The Good Gardener? Nature, Humanity, and the Garden*, London: Artifice, 2015.

Harrison, Robert Pogue. *Gardens: An Essay on the Human Condition*, Chicago: University of Chicago Press, 2008.

Hunt, John Dixon. *The Afterlife of Gardens*, Philadelphia: University of Pennsylvania Press, 2004.

Johnson, W. Ralph. "A Secret Garden: *Georgics* 4.116–148," in David Armstrong, Marilyn Skinner, and Patricia Johnson, eds., *Vergil, Philodemus, and the Augustans*, Austin: University of Texas Press, 2004: 75–83.

Lawrence, Elizabeth. *The Little Bulbs: A Tale of Two Gardens*, Durham, NC: Duke University Press, 1986.

Maddox, Brenda. *Rosalind Franklin: The Dark Lady of DNA*, New York: Harper Collins, 2003.

Moore, Lisa. *Sister Arts: The Erotics of Lesbian Landscapes*, Minneapolis: University of Minnesota Press, 2011.

Parker, Rozsika. "Unnatural History: Women, Gardening, and Femininity," in Noël Kingsbury and Tim Richardson, eds., *Vista: The Culture and Politics of Gardens*, London: Frances Lincoln, 2005: 87–95.

Pugh, Simon. *Garden—Nature—Language*, Manchester: Manchester University Press, 1988.

Ross, Stephanie. *What Gardens Mean*, Chicago: University of Chicago Press, 1998.

Thomas, Richard. "The Old Man Revisited: Memory, Reference, and Genre in Virgil *Georgics* 4.116–48," *Materiali e Discussioni* 29, 1992: 35–70.

Wells, S. *The Dialectics of Representation*, Baltimore, MD: Johns Hopkins University Press, 1985.

1 Garden Design as Feminist Ground

Thaïsa Way

Virginia Woolf famously argued that if a woman were given a room of her own and money "then the opportunity will come and the dead poet who was Shakespeare's sister will put on the body which she has so often laid down."[1] As Elise Smith notes in this volume, Woolf describes the garden "as an opportunity to express a personal vision, explore the process of perception, and provide oblique social commentary."[2] Might the garden serve as Woolf's room of one's own?

For African American women the garden might offer a "last place they thought of," as invoked in the work of Katherine McKittrick.[3] In such a reading the garden offers an essential place in the ongoing geographic struggle of Black women as described by McKittrick as well as Dionne Brand, bell hooks, Alice Walker, and Zora Neale Hurston, alongside historians Diane Glave and Jacqueline Jones among others. The garden might serve as a place within the land and the geography of the material world, "infused with sensations and distinct ways of knowing."[4] It is a place that has been created, cultivated, and enjoyed by Black women who find meaning in the practice of gardening as well as within the place of the garden.[5] In writing about the gardener and poet Anne Spencer, Carlyn Ferrari describes how the garden performs as a place where African American women can "create and cultivate their own space of happiness."[6] It is a place where African American women claimed "a space in which just being and becoming are enough."[7] Such views suggests how gardens offered a form of place-making as an act of "making one's self legible."[8]

Understood as such, a garden as a place of one's own creation, it becomes a liberatory space in which gardeners assigned meaning and found pleasure in being. The garden as such is essential for a woman's creative pursuit and furthermore for her essential experience of being. Indeed, women at the turn of the century transformed the garden from an instrument intended for nourishment to a tool for creating art as well as constructing civic leadership. Gardens in this frame perform essentially, in the words of Greta Gaard, as "alternative conceptions of nature."[9] Taking this farther, the garden is not merely a place for individual claims but a space of collectivity, a place for building community.

This chapter interrogates a broader inquiry into the distinct experiences of overlapping constellations of women in gardens, with a focus on the United States.[10] I draw from feminist histories that center women and questions of gender, while interrogating the intersections of race with a specific eye towards the similar and distinct ways

DOI: 10.4324/9781003381549-2

that the garden was experienced by White and Black women, two constellations for which the garden was an essential place for realizing visions of self and community.[11] The garden is read as a place and an act of subversive agency as women turned what was deemed to be inconsequential space into a place of radical engagement within their communities, collectively pushing the boundaries of gender and race.[12] My intention is to recognize a diversity of women's experiences as they navigated communities and changing assumptions that emerged at the intersections of gender and race while acknowledging that class among other complex forces also shaped world views.[13] The narratives selected suggest how the garden performs as a critical space in which to trace women engaged in creative collaborations, with attention to both the affordances and the constraints of identity and position. A significant distinction is evident in how Black women curated the garden as a place of creative engagement and collective leadership and cooperative engagement in their communities, while for White women the garden more often served initially as a portal to collective leadership and then to the emerging profession of landscape architecture, a more individual pursuit.

The garden in this reading is critically a place of both making and becoming. The garden is environmental, while performing as an aesthetic, cultural, intellectual, and political space of power. As bell hooks writes, "To tend the earth is always then to tend our destiny, our freedom, and our hope."[14] Women in diverse cultures and across time experienced the garden as an intellectual and material place where they have found dignity, beauty, and strength. This is evoked in Alice Walker's description of how "[g]uided by my heritage of a love of beauty and a respect for strength—in search of my mother's garden, I found my own."[15]

Such descriptions suggest how the garden as a work of art arises in the act of gardening, a practice inherently grounded in a collaboration with the natural world. The garden is always becoming as is the gardener within the garden. Donna Haraway proposes that "becoming is always 'becoming with' in a contact zone where the outcome, where who is in the world, is at stake."[16] She argues that the complex entanglements of humans and more-than-human are consistently shaping each other while simultaneously producing knowledge, power, and biophysical material. The garden and gardener are realized by the dynamic forces of nature, culture, and art, in collaboration with one another. The garden and gardener are 'becoming with,' that is the garden is both of and by the gardener and nature. This dynamic character of the garden has been used by women to make a place their own and, equally important, to re-imagine themselves as communities.

The garden acts thus as a space of feminist performance that subsequently functions as an archive for questions of gender, nature, and culture.[17] This chapter focuses on the twentieth century as a period when the garden as a space of "becoming" circumscribed a place in which women might curate alternative identities in the public realms, specifically within the United States. The garden is read as an agent in the negotiations of socially constructed gendered spaces and practices.[18] Furthermore, the garden as a revolutionary act is interrogated as an essential lens for re-reading the history of landscape architecture as well as how Black women suggested alternative forms of collective presence through the claiming of the garden as place.

The Tapestry that is the Garden

The garden is generally thought to be an enclosed and cultivated place; where one dwells and tills; a place that retains a hold on the human imagination.[19] Such a capacious definition may well be why the garden remains such a powerful space and idea. In the vocabulary of Frances Hodgson Burnett's *Secret Garden* (1911), the garden is a place with "Bulbs an' sweet-smellin' things—but mostly roses," [20] offering, as Woolf suggests, an escape "from the common sitting-room [where one can view] … the sky, too, and the trees or whatever it may be in themselves ...."[21] The garden becomes the means for an artist to thrive as both Woolf and Walker describe the human agony of being incapable of nurturing one's creative spirit, a suffering that has led to insanity and perhaps even claims of witchcraft. While Woolf seeks the room to write, Walker looks to her mother in the garden to cultivate her soul describing how her mother tends to her flowers becoming "radiant, almost to the point of being invisible except as Creator; hand and eye. She is involved in work her soul must have."[22] The artist curates the garden as a place where one cultivates creativity drawing on the medium of earth and sky in a continuous process of "becoming with."[23] The garden is a place of deep collaboration, between humans and with the more-than-human world.

The earliest-known garden book for women comes from Great Britain, William Lawson's *The Country Housewife's Garden* (1618). Lawson emphasized the essential nature of a productive garden cultivated by an industrious housewife. Charles Evelyn's *The Ladies' Recreation* (1707) described how women might garden without soiling their hands or challenging their intellectual capacity by focusing on how to lay out lawns and orangeries, and how best to place sculpture.[24] Such ornamental pleasures contrasted with the mundane activities of sowing, planting, and propagating common flowers, knowledge he assumed every "Country Dame" held, but elite women need not pursue.[25] Such gendered views of the garden and gardening would both constrain and afford transgressions as women expanded the boundaries of their knowledge and engagement in the garden and on the land.

Women in the colonies explored an approach more relevant to their circumstances. School teacher, horticulturalist, and gardener, Martha Daniell Logan (1704–1779) of Charleston, South Carolina used the garden as a body of knowledge, writing one of the first American garden treatises, *A Gardener's Kalendar* (1756).[26] She offered practical advice on the cultivation of a kitchen garden as a necessary part of any home and encouraged gardeners to share their seeds and plants, emphasizing the role of collaboration and community.

Garden books by women emerged at the turn of the century as a genre, including Louise Shelton's *The Seasons in a Flower Garden* (1906), Grace Tabor's *Suburban Gardens* (1913), Ruth Dean's *The Liveable House, Its Garden* (1917), and Elsa Rehmann's *American Plants for American Gardens* (1929).[27] Landscape architect, Elizabeth Lawrence (1904–1985), the first published author to explore Southern gardens, designed, gardened, and wrote for a national audience. Magazines including *House Beautiful* and *House & Garden* commissioned women to write about gardens and gardening. By the late nineteenth century, Anna B. Warner would describe the lucrative potential of the garden in *Miss Tiller's Vegetable Garden and the Money She Made by It* (1873). Marion Cran's 1918 book, *The Garden of Ignorance: The Experience of a Woman in a Garden*, told of how she trained her daughter to be able to earn wages from her labors. As landscape historian Dianne Harris has noted, books and essays by women on gardening "contributed to the formation of a network of women's specialized knowledge and support."[28]

Black women's advice on gardening and gardens could be found in club newsletters, however it was most often shared by word of mouth, across generations of women. As bell hooks writes of her community of Black women: "We were indeed a people of the Earth."[29] African American yards and gardens are evident within the quarters for enslaved communities as well as in the Reconstruction era as sites of work and leisure in which traditions of environmental stewardship were cultivated.[30] Recent scholarship reveals how enslaved women modified and interpreted spaces of sustenance, comfort, joy, and sometimes profit as a means of conserving their humanity and their communities.[31] Dianne Glave's *Rooted in the Earth: Reclaiming the African American Environmental Heritage* narrates how the act of transforming yards served as a means for enslaved women to claim authority over their own place, as well as setting such places apart from those of the slaveholders.[32] Additionally, some enslaved women gardening on small plots within the plantation were at times able to sell their produce be it to the enslaver or to others in the community.

After the Civil War, freed Black women used the garden again as a means of making and claiming space, in addition to providing food for their family, as described by Jacqueline Jones in *Labor of Love, Labor of Sorrow*.[33] As with Walker's mother, the garden was a place where she stewarded her identity as woman, mother, friend, and artist. Black women serving as extension agents in the early twentieth century would encourage women to seek profit from their gardens and land. As with women across multiple cultures, gardens were significant places within their lives and communities.[34]

The garden was not an autonomous place, but was integral to the homeplace, whose making hooks describes as one of resistance:

> This task of making homeplace was not simply a matter of black women providing service; it was about the construction of a safe place where black people could affirm one another and by so doing heal many of the wounds inflicted by racist domination. We could not learn to love or respect ourselves in the culture of white supremacy, on the outside; it was there on the inside, in that "homeplace," most often created and kept by black women, that we had the opportunity to grow and develop, to nurture our spirits.[35]

The home, as described by Paula Giddings, "was not so much a refuge from the outside world as a bulwark to secure one's passage through it."[36] Homeplace, inclusive of the garden, as a place of resistance was where one might "regain lost perspective, give life new meaning ... that space where we return for renewal and self-recovery, where we can heal our wounds and become whole."[37]

The garden performed as an interstitial space in which women developed the capacity to move through and where they constructed agency, as individuals and as cooperative communities. It was a place composed of entangled layers of narratives and counter narratives. Such narratives cannot be delinked from one another because the garden is a whole accumulated over time and space. This is evident in the garden's physicality, comprised of earth, sky, and plants, often from different places, emerging in relationship to one another. It is equally descriptive of how the garden offered women a space in which they identified the means to sustain and curate their creative spirit. The garden enabled women to simultaneously stake a place in both the domestic and the civic domains. The garden was both a place of aesthetic pleasure, individual transgression, and collective revolution.

The garden as a site of becoming can be easily traced in the garden of Harlem Renaissance poet, gardener, librarian, and activist Anne Bethel Spencer (1882–1975) in Lynchburg, Virginia. While Spencer cultivated her life as an artist, she was also an activist working to resist Virginia's Jim Crow laws and the gendered assumptions about women as wives, mothers, and writers including by helping to found of the Lynchburg Chapter of the National Association for the Advancement of Colored People.[38] Her home served her purposes well. Spencer's home and garden can be described as her homeplace and a gathering place for friends, specifically Harlem literati including Langston Hughes, W. E. B. DuBois, Zora Neal

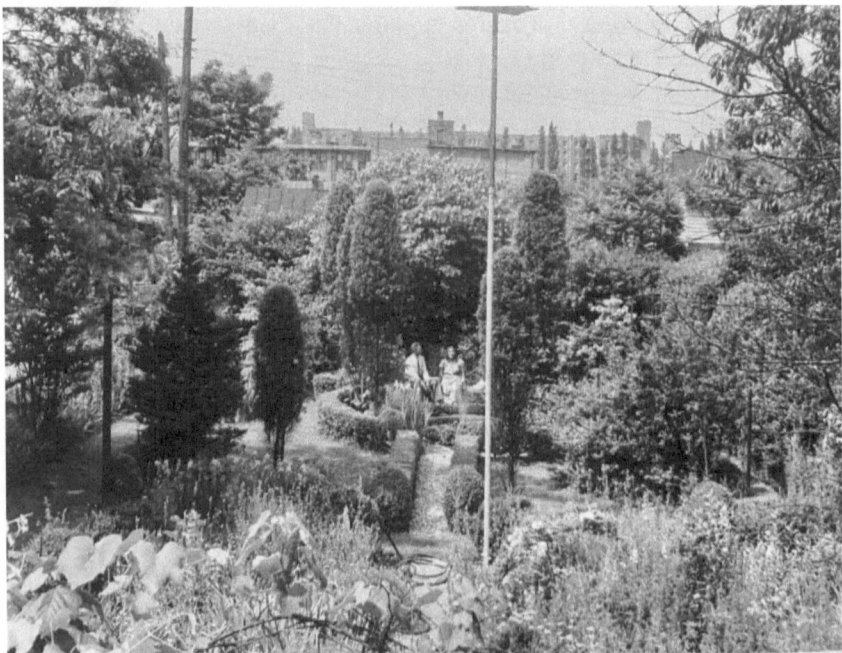

Figure 1.1 Anne and Edward Spencer in the garden, Lynchburg, VA. Anne Spencer House and Garden Museum.

Hurston, Mary McLeod Bethune, Gwendolyn Brooks, and James Weldon Johnson, who published her first poem in *The Crisis.*

Moving to the Queen Anne style home built by her husband Edward in 1902, Spencer first planted a vegetable garden. As their children grew, she reimagined her garden as a work of art. Spencer, whose father was born into enslavement and mother a descendant of enslaved people, named her garden with its writing cottage Edankraal, combining her and Edward's names, and *kraal*, the Afrikaans word for dwelling.[39] She designed, planted, and tended to her garden in partnership with Edward who built the structures including the pergola, arbor, pond, and her writing studio.[40] The garden would become a "fusion of recycled architectural elements and high art garden designs."[41] As Ferrari notes, Spencer's garden was a site of "self-fashioned, self-defined pleasure and artistic inspiration, but more importantly, her garden figured her Black womanhood."[42] While the garden might appear at first glance to resemble a colonial revival garden, Spencer subverted such a trope by cultivating plants collected on trips to the countryside and exotic plants given her as gifts. Taking a special place was the cast iron head of an African prince, a work of the Ebo tribe of West Africa, a gift from W. E. B. DuBois. The abundance of flowers and brightly colored materials comprising gates, arbors, and garden furniture, as well as the creative use of recycled materials, may be more typical of a Southern rural garden as evoked by Walker's descriptions, and by Vaughn Sills' photographs (in this volume). Spencer's garden performed as a private retreat for her and a shared community space, a place of artistic and horticultural creation, a muse for her poetry and art, and a place for the intellectual engagements of a community of artists, writers, and activists of the Harlem Renaissance. As a place of cooperative and collective engagement, the Spencers used the garden place to nurture their family and community. The garden was not only a reflection of individual creativity, but a place of collective nurturing of an African American community with and in the land.

Described by Ferrari[43] as a feminist interested in womanist themes, Spencer's artistic production, both her poetry and her garden, suggest an intimately woven tapestry of poetry and gardens tended to over a lifetime.[44] Gardening and writing poetry and prose, nurtured one another. The garden was where Spencer wrote much of her poetry; she drew deeply from the garden literally in her poetry. Her poems depict her garden as an essential space of inspiration and beauty, a sacred space and a retreat, and a metaphor for transformation, wounding, and healing.[45] This character of the garden as more than the individual gardener, with a future extending beyond the artist is reflected in her poems. In "Any Wife to Any Husband," she writes:

> This small garden is half my world
> I am nothing to it when all is said,
> I plant the thorn and kiss the rose
> But they will grow when I am dead.
>
> Let not this change,
> Love, the human life

Share with her the joy you had with me,
List her with the plaintive bird you heard with me,
Feel all human joys, but
Feel most a "shadowy third."[46]

As a wellspring for her creative production, Spencer reworked her garden as she reworked her poetry. Her garden and her poetry reflect her engagement with contemporary discourses as well as her commitment to protesting inequalities deriving from racism and sexism through the act of nurturing a collective community. Her garden was physically and poetically layered with meaning, experience, and metaphors. It was imbued with the power to nurture her creative and aesthetic spirit as well as her political and cultural visions of a more just community in partnership with the natural world.

Civic Improvement and Transformational Leadership

As evident in Spencer's art, the garden served as a stage for transgressive performance of the gardener and her collaborators at the same time as it performed as a revolutionary space. This dual nature would challenge the gendered and racial assumptions about women in the public realm. Women's place was assumed to be within the home, the civic and public realms were assigned to men. The garden as in-between space would be the site for women to stake a place in the domestic and the civic spheres and to take part in radical acts to create a more just world.[47]

The nineteenth-century's village improvement movement reveals the porosity of the domestic and the civic in women's leadership. Led by Mary Hopkins Goodrich (1814–1895) and the Laurel Hill Association, and formed for the purpose of beautifying Stockbridge, Massachusetts in 1853, the goal was described as "to improve and ornament the streets and public grounds of the village … and generally doing whatever may tend to the improvement of the village as a place of residence."[48] During this Progressive Era, as guardians of virtue and morality, women were considered naturally associated with labors to cleanse and purify the municipal, state, and federal houses and to upgrade the quality of life for their brother and sister Americans. Jane Addams (1860–1935), leader of the Settlement House Movement, advocated for women and children to have access to healthy places in which to live and work. She promoted public engagement with issues of concern to women and more broadly community public health, and peace around the world. Addam's 1907 essay "Utilization of Women in City Government," [49] describes the intersecting concerns of good government and good housekeeping that framed the efforts as appropriate for women's leadership. The authority granted women in the domain of housekeeping and moral values provided, if unintentionally, a visible place in the public sphere for women, collectively and individually.[50] As White women constructed their professional roles as landscape architects, they drew on the legacy of village improvement principles in their design of model gardens intended to "teach" taste to the masses.

Annette Hoyt Flanders (1887–1946) designed the gardens for America's Little House in New York City. Although known for country estates, her lectures, publications, and exhibition gardens promoted her vision of America as a nation of small gardens where each gardener could "Make a more permanent contribution to the cause of beauty."[51] She sought to teach taste to middle-class homeowners by illustrating the "logical sequence in which things should be done to develop property practically, beautifully, and economically."[52]

Better Homes in America Campaign, first launched in 1922 by the President's Advisory Council on Housing, sponsored model homes throughout the United States. Flanders's garden was visited by over 145,000 guests who were provided with a budget illustrating the affordability of such a modern garden. The garden was described by Pearl Buck as an "oasis among the towering skyscrapers, ... [created by placing] simply a layer of topsoil ... two feet thick ... over discarded bricks, broken bottles and other rubbish."[53] This description suggests how the garden might transform the detritus of the city, even an abandoned plot, into a place of beauty. The colonial revival garden and house stood in contrast to the city, in scale, style, and texture. The house and garden were imagined as countering urban blight, a perspective that echoed village improvement advocates as they reimagined the city as collections of well-tended gardens. Equally important is how the garden suggested a specific White-colonial affiliation as it was designed in the popular colonial revival style known from Colonial Williamsburg, Virginia. The colonial

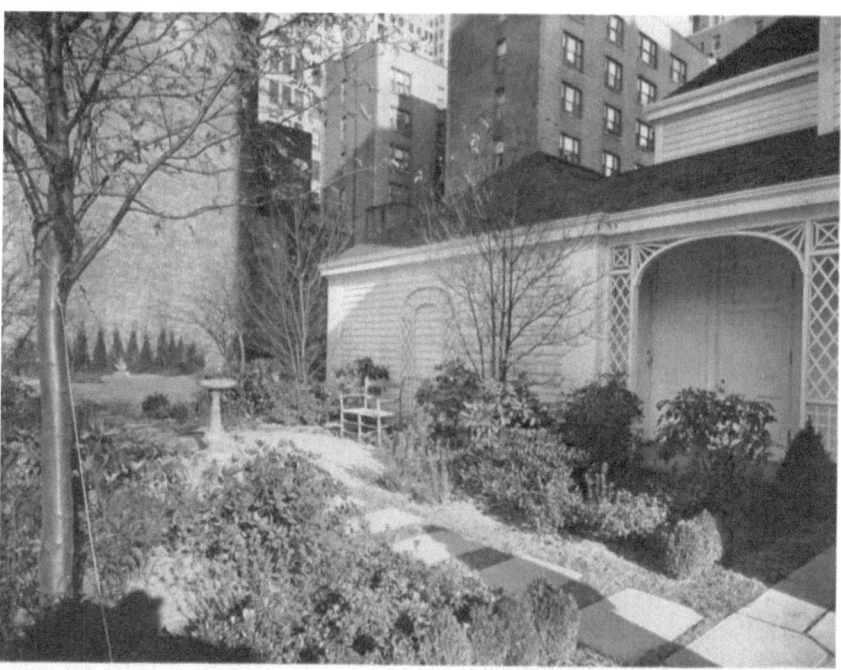

Figure 1.2 America's Little House with garden designed by Annette Hoyt Flanders. America's Little House, *Better Homes in America*, Photograph by Bert Lawson, 1934, p. 20.

revival style spoke to the predominantly White audience actively engaged in literally white washing the history of the colonial era of North America. Flanders' work both as a designer and in her essays for popular magazines emphasized the capacity of the average American to design a garden that would improve their daily lives, and in turn, their community, while promoting a racial and class-based identity.

Garden Clubs

Garden clubs emerged in the early twentieth century as powerful models of women's collaborative approach to leadership and the increased fluidity with which women moved between the realms of private and public. The women's club movement offered women an independent avenue for self-improvement, education, and civic engagement, specifically and most notably in the women's suffrage movement. While suffrage served as a contested topic, women's active participation in garden clubs was viewed as socially appropriate. The earliest garden club in the United States was created in 1891 as the *Ladies' Garden Club of Athens, Georgia*. By 1913, White women organized a national federation of garden clubs, the Garden Club of America, and in 1929, the *National Council of State Garden Clubs*. These garden clubs had three main purposes: to strengthen woman's role as protector of domestic life and defender of values, morals, and patriotism; to preserve the nation's history; and to improve the standards of gardens, both private and public, throughout the nation. As noted in 1913 in *House Beautiful*,

> From our own grounds the next logical step is to the attention to parkways and streets and from these again to membership on park boards and civic improvement leagues .... The woman who steps from her doorstep to plant a few handy 'yarbs' at its foot, has quite unwittingly taken her first unconscious step towards the suffrage .... Most of us do not realize as we wield trowel and rake that we are doing a deed of national significance.[54]

Martha Brookes Brown Hutcheson (1871–1959) was a professional landscape architect and member of the American Society of Landscape Architects (ASLA), but her legacy, as landscape historian Roxi Thoren has written, was built upon her collaborations with the Garden Club of America as a platform for her social agenda.[55] After studying landscape architecture at Massachusetts Institute of Technology (MIT), Hutcheson opened her office in 1902, and, until 1910, maintained a firm based in Boston and New York. After 1910, she focused on writing and lecturing, returning to the profession at the end of her career. She was a founding member in 1913 of the Somerset Hills Garden Club, an affiliate of the Garden Club of America. Hutcheson's intent in pursuing a career was to "bring about positive social changes through landscape design."[56] She believed that gardens and gardening were essential social and cultural contributions by women and that it was important to identify the means to teach women modern techniques and skills in the garden. During World War I, Hutcheson invited the Woman's Land Army of America to her farm to learn improved agricultural techniques and practices.

Hutcheson considered landscape design a fundamental force for civic betterment and encouraged the clubs to open their meetings to a broader audience so as to reach a greater community of women.[57]

As models of civic responsibility, Hutcheson encouraged garden club women to create "the finest gardens which America can produce" to demonstrate better standards for the public. She designed her landscape at Merchiston Farms as a demonstration garden. Her book, *The Spirit of the Garden*, dedicated to "those with a progressive spirit in their concern for the fine art of garden making," was intended to teach "good taste."[58] As she wrote in a manifesto for the Garden Clubs to broaden their programs, women must "Carry on! To our country's beautifying, and begin first with our neglected rural towns and let our generation start the great ball rolling, for our opportunity is great."[59]

The native plant movement was amplified during Hutcheson's career at the same time as the increasing national concern about immigrant communities. Hutcheson promoted the use of local plants in appropriate habitats both for ecological reasons and to create economically feasible gardens for working-class families, revealing a proto-ecofeminist framework.[60] She advocated the use of native plants acquired locally, and shared, in contrast to exotics that must be purchased and often did not thrive in new environments. She wrote that gardening with local plants would foster assimilation of immigrants into the native landscape and nation. While she appreciated the gardens featured at club events, she argued they were too expensive for most families. Instead, members, she wrote, should demonstrate ecologically appropriate gardens, that would also be affordable and more easily emulated by a broader community of households.[61] Hutcheson framed her design approach to consider gardens as simultaneously aesthetically appropriate and community-oriented, modeling how women might lead with a social agenda, although she remained quiet on questions of race and immigration.

Black women have historically been deeply engaged in cooperatives and collective actions including during the Civil War when African American women formed the Combahee River Colony in South Carolina to grow cotton on abandoned farms.[62] African American women formed clubs to facilitate mutual aid in support of their communities and to empower their sisters as engaged and informed citizens.[63] One of the earliest African American women's clubs was the Female Benevolent Society of St. Thomas, in Philadelphia, founded in 1793, in order to "secure shoes and clothing for women and children who had been rescued by the Underground Railroad."[64] African American women organized throughout the nineteenth century, often through churches.[65] The National Association for Colored Women's Clubs was founded in 1896 in Washington, DC by members of the National League of Colored Women, which was founded in 1892 in Washington, DC, and the National Federation of Afro-American Women, formed in Boston during the First National Conference of Colored Women of America in 1895. Their shared agenda was explicitly focused on political, social, and community improvements. Under the leadership of women such as Mary Church Terrell (1863–1954) they sought progress with the motto "Lifting as We Climb."[66] Margaret Murray Washington (1865–1925) served in many roles expanding her primary role as Lady Principal at

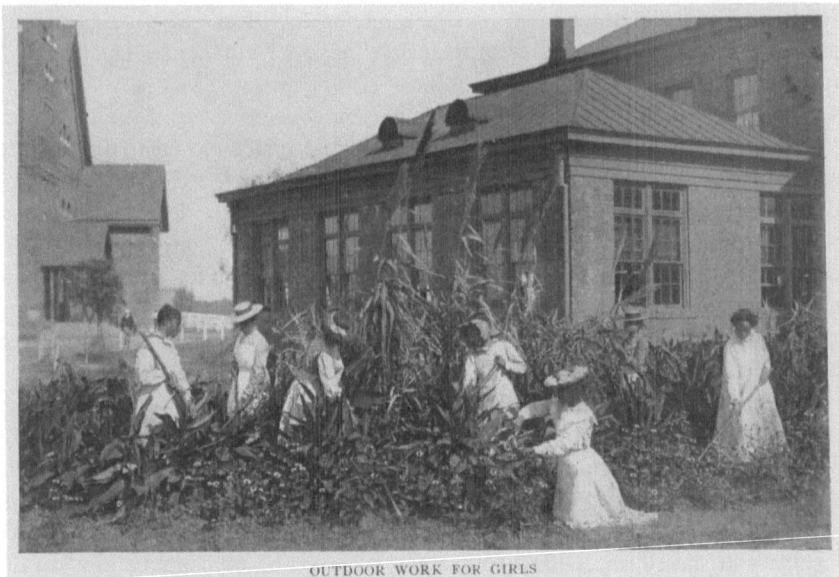

OUTDOOR WORK FOR GIRLS

Figure 1.3 "Outdoor Work for Girls." Women students gardening at the Tuskegee Institute. Schomburg Center for Research in Black Culture, Jean Blackwell Hutson Research and Reference Division, The New York Public Library. "Outdoor Work for Girls," *The New York Public Library Digital Collections*, 1904.

the Tuskegee Institute in Alabama. She was instrumental in the Tuskegee Women's Club, the National Federation of Afro-American Women's Clubs, the National Association of Colored Women, the Southern Federation of Colored Women, the Alabama State Federation, and the International Council of Women of the Darker Races of the World.[67] Such clubs sought to uplift the masses and to improve the lives of all African American women and families, an approach that differed from the clubs for White women that most often focused on middle-class women and their opportunities.[68]

Participation in the Negro garden clubs contributed to a variety of pursuits that also drew from a legacy of farming and agricultural practices.[69] African American women served as early Home Demonstration Agents for the US Department of Agriculture including Annie Peters (dates not found), the first African American woman appointed as a federal agent in Oklahoma on January 1, 1912, with Mattie Holmes (dates not found) hired a few months later in Virginia.[70] The agents were assigned to advise rural families on how to improve their homes through advancements in the community, the field, and the garden.[71] Through the garden and environmental stewardship practices, the women established leadership roles in improving their communities through collective cooperation. These efforts would contribute to an emerging community of Black cooperatives that included women leaders such as Ella Baker (1903–1996) who helped found the Young Negroes' Cooperative League in 1930. Decades later in 1969 Fannie Lou Hamer would launch the Freedom Farm Cooperative for farmers, a critically important collective project.[72]

Club members focused on education, health, and community improvements while developing organizing and leadership skills.[73] An emphasis on the built environment is evident in the work of the Detroit Gardens Council, organized in 1924 as a branch of the Woman's City Council, whose first project was to improve the Eight Mile Road district of Detroit.[74] In 1931, a community of Black women in Dayton, Ohio founded the Delphinium Garden Club led by Remitha Ford (dates not found), a local social worker.[75] In Indianapolis, Black women formed their own Delphinium Garden Club in 1938 that remained active for thirty-five years. A 1963 history of the club shared its commitment "to develop genuine appreciation for the healing power of nature's bounty and beauty in a perplexed world."[76] The *Negro Garden Clubs of Virginia* was founded in 1932 with seven chapters expanding within a decade to sixty-five chapters. The club members promoted the garden as a creative work, a place of refuge, and a source of sustenance, as well as the benefits of gardening as an activity and body of knowledge.

Garden club members organized activities including workshops, tours, and book clubs to share knowledge on garden plants, design, and horticulture, as well as community building. Nature study, school gardening, and the formation of garden clubs were intended to improve the social and economic development of rural areas. For some, including the Club of Virginia, the focus of these organizations extended to the purview of "improving race relations,"[77] suggesting the strong connections between the club missions and efforts to be recognized as legitimate citizens and communities of the nation. The *Negro Garden Clubs Handbook*'s essay, "Ten Years of Progress by the Negro Garden Clubs of Virginia," noted that "We have succeeded in getting the majority of Negroes to beautify their homes and yards. At the same time, these citizens have become civic-conscious and have become voters."[78] Although by the 1970s garden clubs were less popular, the women remained engaged in civic leadership, many leading collective fights against urban renewal and in support of community improvements in their own neighborhoods.[79]

A focus on civic engagement was typical of the organizations formed by African American women.[80] Gardens offered women an instrument of engagement, leadership, and eventually radical change. The garden was much more than a plot of tilled ground as it served to provide a liminal space between the private sphere and the public sphere. Decades later, Black women would enter the profession of landscape architecture.

A Profession of Landscape Architecture

In the early twentieth century, White women were able to leverage their engagement in the garden into the profession of landscape architecture reflecting an emphasis on individual success. The "new" White women built upon the cultural tropes of feminine areas of expertise—including gardens, gardening, horticulture, decorating, and design— as well as their White privilege to claim new authority as professionals.[81]

While the earliest recognized work of landscape designers were public cemeteries and parks, the profession soon embraced a broader definition of practice. As noted in the weekly journal, *Garden & Forest*, landscape architecture was considered a "broad and catholic art... as useful in the preservation of the Yosemite Valley or the scenery of Niagara as it is in planning a pastoral park or the grounds about a country house."[82] Art critic Mariana Griswold Van Rensselaer wrote a series on "Landscape Gardening," [83] describing landscape architecture as a new title for the landscape gardener, as a fine art warranting the respect given painting, sculpture, and architecture.[84] This definition of the landscape architect aligned the practice with theories of art and aesthetics rather than trades or crafts.

With increased public recognition and new educational programs, the profession was made official in 1899 with the founding of the American Society of Landscape Architects (ASLA) by a group of White male and female practitioners. Of the founding practitioners, five studied horticulture and/or worked in nurseries, two were trained as park designers, three were educated in civil engineering, and one was an architect. Beatrix Jones (later Farrand) was a founding member. Elizabeth Bullard from Connecticut was elected to the association a few months later. With knowledge in horticulture, botany, gardening, and the fine arts considered socially appropriate, White women could, with relative ease, transition from a role as a garden designer, and even gardener, into a professional role as a landscape architect.

Early professionals had studied by enrolling in courses in landscape and garden design and gardening offered as early as the 1870s as part of the land-grant college curriculum. By the early twentieth century, coeducational programs were being established at Cornell, University of Illinois, University of Iowa, Michigan State, and MIT. Lowthorpe School of Landscape Architecture for Women opened in 1901 just as the Harvard program was being launched. The Pennsylvania School of Horticulture for Women was opened in 1910, and the Cambridge School of Architecture and Landscape Architecture for Women in 1915. Their programs taught the art and science of designing gardens, landscapes, parks, and city planning as preparation for women to become landscape architects.

Similar courses were available for Black women starting in the 1890s as the land grant program expanded to support African American colleges (historically Black Colleges and Universities/ HBCUs).[85] These efforts were contemporaneous with increased access for Black women to attend colleges, where by 1935 more women were enrolled than men.[86] Initially courses were offered for men in departments such as Mechanical Industry and Agriculture at schools such as the Hampton Institute where students could enroll in courses in horticulture, agriculture, and landscape gardening. The Tuskegee Institute developed the courses specifically for women after Margaret Murray Washington and her husband Booker T. Washington made a trip to England in 1899, when they visited the Swanley Horticultural College.[87] While Swanley originally enrolled male students, by 1894 female students comprised the majority and in 1903 it became a women-only institution.[88] Observing women learning to garden as an art and a science served as a model for Mrs. Washington's commitment to strengthen and expand the curriculum at Tuskegee.[89] David Williston, the first professional African American landscape architect, was hired in 1903 as professor of horticulture and then as campus landscape architect. He would teach at the school for twenty-seven years.

Despite the opportunities for education offered to Black women, the emerging profession of landscape architecture was limited primarily to White practitioners. A 1928 report on women in architecture and landscape architecture noted that such professions were "peculiarly adapted to women … [and] appeal to the naturally artistic feminine instinct."[90] The report further contended that women "have a natural aptitude for the more intimate type of design that one finds in domestic work, coupled with an instinct for plant design and groupings."[91] Beatrix Farrand and Ellen Shipman offer two narratives of how White women pursued professional practice. Both excelled in the design of gardens as exquisite works of art that relied on extensive horticultural knowledge and high levels of stewardship. Such women reflect a broader constellation of White women who used the garden as a bridge to a professional practice in ways that have come to define the profession of landscape architecture today.

Beatrix Jones (later Farrand) (1872–1959) was born to "five generations of garden lovers."[92] Her mother, Mary Cadwalader Rawle, was a member of a community of artistic minds serving as literary agent for her sister-in-law, Edith Wharton. Educated by tutors in the arts and humanities, Beatrix sought out Charles Sprague Sargent, curator of Harvard's Arnold Arboretum with whom she studied botany, horticulture, and arboriculture. She had a tutor to learn engineering. She launched her design practice in 1896 in New York City. While her repertoire included botanical gardens, university campuses, and public gardens, her most recognized designs

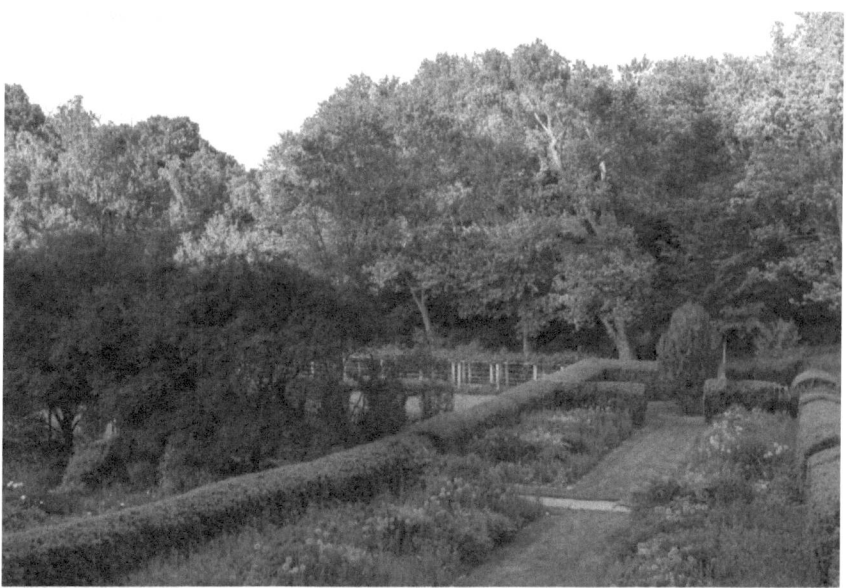

Figure 1.4 Herbaceous Garden designed by Beatrix Farrand, Dumbarton Oaks. Photograph by Sahar Coston Hardy for Dumbarton Oaks.

are Dumbarton Oaks for Robert and Mildred Bliss in Washington DC and the Eyrie for Abby Aldrich Rockefeller in Seal Harbor, Maine.

Farrand determined early in her career to call herself a landscape gardener suggesting the significance that she placed on the practice of design as one centered on gardens and gardening. Although it may be that she chose the title to denote the lack of a professional education, her use of "landscape gardener" suggests how Farrand privileged the art of gardening in the practice of landscape architecture. She objected to the use of the word architecture in the title as inappropriate to the practice.[93] Additionally, as an anglophile, Farrand may have appreciated the title's picturesque connotations. In her determination *landscape gardener* was the best description of her practice and design philosophy. This approach to design was nowhere more evident than in her design of the Dumbarton Oaks gardens and landscape.

Purchased by Robert and Mildred Bliss in 1920, the site included a house and a series of gardens as well as a collection of trees and shrubs planted and cultivated by enslaved labor at the direction of Edward M. Linthicum who owned the property from 1846 to his death in 1869. The Bliss couple commissioned Farrand to design their country estate, a project that would extend over twenty years. From 1921 to 1947 Farrand oversaw the design of the fifty-three-acre landscape including a ten-acre formal garden, documented in over 1,226 drawings and sketches, and described in her *Plant Book*.[94] Farrand partnered closely with her patron and "Gardening Twin" Mildred Bliss, consulting on every detail at every scale as the gardens were developed.[95] Drawing from the knowledge of both women, the landscape exhibits European influences while remaining an American work of art reflected in how the design fits the ground and the patron's lives through a weaving together of formal and informal designs. The fitting of the design stems from Farrand's capacity to design as an artist while building on her knowledge of topographic engineering alongside her horticultural expertise that supported Bliss's vision of a garden.

A different narrative is evident in the career of Ellen Biddle Shipman (1869–1950), described as "one of the best, if not the very best, Flower Garden Maker in America."[96] In 1917 *House and Garden* named her the "dean of American women landscape architects," recognizing her role as a teacher, leader, and professional.[97] Having established a practice in Cornish, New Hampshire, Shipman opened her office in New York City in the early 1920s.[98]

Shipman believed that beautiful landscapes and gardens were essential for the nation as well as individuals' intellectual and spiritual growth. For Shipman, gardening was

> an art of much more vital importance than most Americans realize.... open[ing] a wider door than any other of the arts—all Mankind can walk through, rich and poor, high and low, talented and untalented. It has no distinctions, all are welcome.

For her, a house was a home when a garden was created, and it was there that democratic values were nurtured. "If one can gauge the height of civilization by the

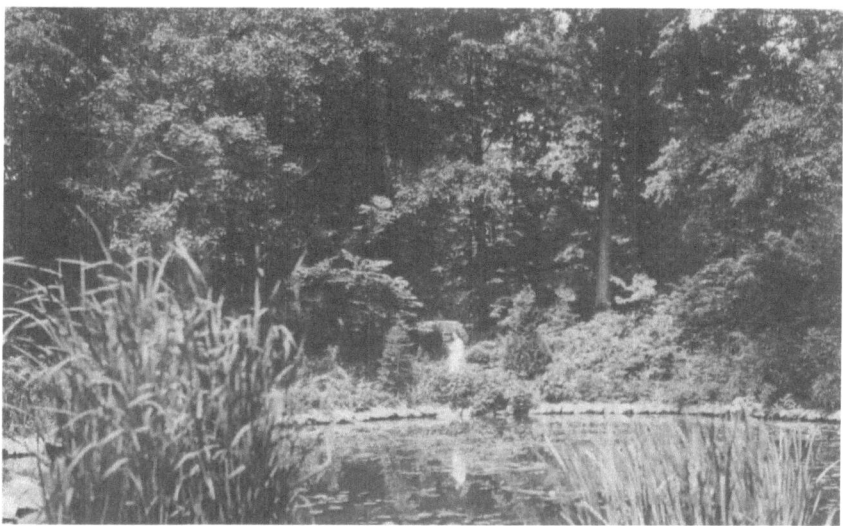

Figure 1.5 The Causeway's wild garden and pond as designed by Ellen Shipman. *Ellen McGowan Biddle Shipman Papers, #1259.* Division of Rare and Manuscript Collections, Cornell University Library, n.d.

beauty of the gardens, one can also judge the spirit of democracy in a people by the prevalence of gardens among all its classes."[99]

An early project for Shipman was the Causeway estate in Washington, DC, designed as a wild garden and woodland within a formal landscape.[100] In collaboration with Charles A. Platt between 1912 and 1914, Shipman designed the landscape to transition from a series of formal flower gardens near the house, to a broad lawn embowered by a native woodland. She preserved existing trees judiciously adding shrubs, flowering plants, and ground covers. The wild garden within the woodland included a small pond, a meandering bridle, and walking paths, and sweeping masses of flowering plants as well as substantial trees defining the canopy. Shipman's wild garden offered new possibilities for garden design, one that became increasingly popular with the native plant movement. As with Dumbarton Oaks and Edenkrall, it was a garden to be actively experienced, through walking, sitting, and gardening.

And More … Until There Were Fewer

Despite increasing recognition for landscape architecture at mid-century, to the frustration of those in the ASLA, the public continued to conflate the profession of landscape architecture and the craft of gardening. Landscape architects sought to disengage their practice from gardening and horticulture to align with architecture and engineering. In the process, practitioners were erased from the canon. Garden designers and artists such as Anne Spencer were ignored, as were landscape architects Ruth Dean, Elizabeth Bullard, and Marian Cruger Coffin. Those who had created a garden of their own found that while it still nurtured the creative spirit, it no longer served as a generative place for transgressions into the public realm.

The gardens designed by women reveal both shared and distinct constellations of artists shaping early practices in landscape design. The narratives suggest the richness of the ways women bridged the craft of gardening with a place in the public realm and community leadership. While White women were able to move into the profession of landscape architecture by the early twentieth century, it was only later possible for women of color and immigrant women. In 1953 after being interned at a war camp during World War II, Mai Arbegast was the first Japanese American woman to complete an MLA from the University of California, Berkeley, where she would later teach until 1966.[101] While African American landscape architect David Williston had established his practice in1898,[102] Karen Phillips, graduating with a B.L.A. from the University of Georgia in 1975, would be the first Black woman to be elected as an American Society of Landscape Architects Fellow only in 1998, nearly a century after the founding of ASLA.[103] While gardens connected private and public realms for women, access to the professions was limited by the racism at the center of professionalism in the United States.

Black women understood how "Reclaiming our history, our relationship to nature, to farming in America, and proclaiming the humanizing restorative of living in harmony with nature so that the earth can be our witness is meaningful resistance."[104] This resistance manifested for many women in new roles and futures as artists, scientists, teachers, writers, and designers in ways that we have yet to fully acknowledge. The legacies of these women's contributions to place and culture through collective stories, lineages, and networks of women as gardeners and garden designers expands the histories of the built environments through questions of race, gender, and identity. The garden of one's own where women nurtured their creative spirits by creating a work of art significantly shaped the profession of landscape architecture. At the same time the garden as a place of collective community work as practiced by Black women offers new ways to think about design and the garden, now and in the future.

Notes

1 Woolf, *A Room of One's Own*, 95.
2 Smith, "Virginia Woolf and Vanessa Bell at Kew Gardens," in this volume.
3 McKittrick, *Demonic Grounds*, 37, 42.
4 McKittrick, *Demonic Grounds*, ix.
5 I have moved between using Black and African American as historically these have been in use with different meanings and connotations. While Black women is common today, for individuals such as Anne Spencer, African American would have been her chosen description according to her grand-daughter, Shaun Spencer.
6 Ferrari, *Do Not Separate Her*, 49.
7 Ferrari, *Do Not Separate Her,* 24.
8 Ferrari, *Do Not Separate Her,* 18.
9 Gaard, "Nature, Gender, Sexuality," 265.
10 I use "constellations" to describe "a specific milieu, at once local, national, and transnational ... [drawing] the term from Walter Benjamin and Theodor Adorno to suggest elements (or people) at once juxtaposed and changing; a definite pattern unit[ing] them

but it overlaps with other patterns and has no inherent or totalizing essence." Jay, *Adorno*, 14–15.
11 From Crenshaw, "Demarginalizing the Intersection of Race and Sex," 57–80.
12 See White, *Freedom Farmers*, 3–27, 65–87.
13 Drawing from "The Combahee River Collective Statement (1977)," 292.
14 hooks, *Belonging*, 68.
15 Walker, *In Search of Our Mothers' Gardens*, 243.
16 Haraway, *When Species Meet*, 244.
17 Parker, "Unnatural History: Women, Gardening, and Femininity."
18 Drawing from Spencer-Wood and Baugher, "Introduction to the Historical Archaeology of Powered Cultural Landscapes," 463–474.
19 Giesecke and Jacobs. *The Good Gardener?*
20 Burnett and Tudor, *The Secret Garden*, 117.
21 Woolf, *A Room of One's Own*, 95.
22 Walker, *In Search of Our Mother's Gardens*, 241.
23 For references to the concepts of "becoming" and "becoming with" see Deleuze and Guattari, *A Thousand Plateaus: Capitalism and Schizophrenia*; Haraway, *Simians, Cyborgs, and Women: The Reinvention of Nature*, 91.
24 Evelyn, *The Lady's Recreation*.
25 Bell, "Women Create Gardens in Male Landscapes."
26 Logan, *A Gardener's Kalendar*.
27 As these are identified within the text we have not included in endnotes or the bibliography.
28 Dianne Harris, "Women as Gardeners," 1448.
29 hooks, "Earthbound: On Solid Ground," 118.
30 See Heath and Bennett, "'The Little Spots Allow'd Them'"; Gundaker and Cowan, *Keep Your Head to the Sky*; Mullins, "Gardens in the Black City"; Glave, *Rooted in the Earth*.
31 Glave, *Rooted in the Earth*, 125.
32 Glave, *Rooted in the Earth*, 117.
33 Jones, *Labor of Love, Labor of Sorrow*.
34 Waldenberger, "Barrio Gardens," 232.
35 hooks, *Yearning*, 42.
36 Giddings. *When and Where I Enter*, 100.
37 hooks, *Yearning*, 49.
38 Ford, "Flowering a Feminist Garden," 8.
39 Frischkorn, Spencer, and Rainey. *Half My World*, 32.
40 Spencer has been rediscovered as part of what Alice Walker describes as the "vibrant and creative spirit that the Black woman has inherited" from her mother and grandmothers. Walker, *In Search of Our Mother's Gardens*, 239.
41 Frischkorn, Spencer, and Rainey, *Half My World*, 27.
42 Ferrari, *Do Not Separate*, 19.
43 Frischkorn, Spencer, and Rainey, *Half My World*; Greene, *Time's Unfading Garden*, 129–47.
44 Ford, "Flowering a Feminist Garden," 11.
45 Frischkorn, Spencer, and Rainey, *Half My World*, 57–90.
46 Greene, *Time's Unfading Garden*, 186.
47 See McKay, *Radical Gardening*.
48 Waring, *Village Improvements and Farm Villages*.
49 Addams, Newer Ideals of Peace, 180–208.
50 Scott, *Natural Allies*.
51 Henry, "Notes", 123.
52 Henry, "Notes", 123.
53 "Little House Gets Crop of 36 Apples," *New York Times* (July 20, 1935), 15.

54 McKittrick, *Demonic Grounds*, ix.
55 Thoren, "Dreaming True"; Davidson, *Spirit*, "Introduction".
56 Davidson, *Spirit*, xxi.
57 Davidson, *Spirit*, xxix.
58 Hutcheson, *The Spirit of the Garden*; Hutcheson, "A Wider Program for Garden Clubs," dedication page.
59 Hutcheson, "A Wider Program for Garden Clubs," 14.
60 Thoren, "Dreaming True". Other advocates for native plants include Jens Jensen, Ruth Dean, and Elsa Rehmann.
61 Way, "Social Agendas of Early Women Landscape Architects."
62 Gordon Nembhard, *Collective Courage*, 38–39.
63 Lerner, "Early Community Work of Black Club Women."
64 Davis, *Lifting As They Climb*, 7.
65 Jones, *Vanguard*.
66 Davis, "Lifting As They Climb,"1–12.
67 Harris, *Margaret Murray Washington*.
68 Giddings. *When and Where I Enter*, 98.
69 Roberts, "The Farmers' Improvement Society." Note that the term "Negro" was used at the time.
70 Scholl, "Extension Family and Consumer Sciences"; *Home Demonstration Agent;* AIB 38-July 1951, US Department of Agriculture, booklet, 25–27.
71 Harris, "The South Carolina Home in Black and White," 477–501.
72 Blain, *Until I Am Free*, 137–141.
73 Scott, "Most Invisible of All"; "Black Women's Club Movement."
74 Wolcott, *Remaking Respectability*, 155–156.
75 "History" *Guide to the Delphinium Garden Club of Dayton Records*.
76 Mullins, "Gardens in the Black City."
77 Williams, *Handbook ,* 84.
78 Williams, *Handbook,* 86.
79 Mullins, "Gardens in the Black City."
80 See among others Jones, *Vanguard*; Roberts and Butler. "Contending with the Palimpsest."
81 There remains a significant gap in research on how non-White women negotiated the opportunities of the garden.
82 Charles and William A. Sprague, "Cover Page", 192.
83 Van Rensselaer, "Landscape Gardening—a Definition," 2.
84 Van Rensselaer, *Art Out-of-Doors*, 3.
85 Neyland and Fahm, *Historically Black Land-Grant Institutions*, 17–77. Today there are four HBCUs with degree programs in landscape architecture.
86 Allen et al. "Historically Black Colleges and Universities".
87 Washington, *Working with the Hands*, 107–118.
88 Opitz, "'A Triumph of Brains over Brute'."
89 Lee, "Profiles in Botany: Margaret James Murray Washington."
90 Frost and Sears, *Women in Architecture and Landscape Architecture*, 24.
91 Frost and Sears, *Women in Architecture and Landscape Architecture*, 24.
92 Farrand, *The Bulletins of Reef Point Gardens*.
93 Begg, "Influential Friends," 41–42.
94 Farrand, *Beatrix Farrand's Plant Book for Dumbarton Oaks*.
95 Way and Coston-Hardy, *Garden as Art*.
96 "W.H. Manning to Frank Seiberling, Akron, Ohio, July 20, 1917," Stan Hywett Archives.
97 Wright, House & Gardens Own Hall of Fame," 50.
98 Way, *Unbounded Practice*; Way "Longue Vue Gardens and Landscape," 192–221.
99 Shipman, "Garden-Notebook."

100 Way and Callcott, "Expanding Histories/Expanding Preservation: The Wild Garden as Designed Landscape," *Preservation Education & Research Two* (2009): 53–64.
101 Asa Hanamoto and Masao Kinoshita were Japanese American men who graduated in landscape architecture in the early 1950s.
102 In 1973 a survey was completed by the American Society of Landscape Architects identifying seven Black landscape architects all of whom were men.
103 See Karen Phillips and Perry Howard, "2020 Landscape Architecture Speaker Series – Karen Phillips and Perry Howard."
104 hooks, "Earthbound: On Solid Ground," 119.

References

Addams, Jane. *Newer Ideals of Peace*, Citizen's Library of Economics, Politics, and Sociology New York: Macmillan, 1911. https://hdl.handle.net/2027/hvd.HW3Q8H: 180–208. Accessed July 17, 2023.

Allen, Walter R., Joseph O. Jewell, Kimberly A. Griffin, and De'Sha S. Wolf. "Historically Black Colleges and Universities: Honoring the Past, Engaging the Present, Touching the Future." *The Journal of Negro Education* 76, no. 3 (2007): 263–280.

Begg, Virginia Lopez. "Influential Friends: Charles Sprague Sargent and Louisa Yeomans King." *Journal of the New England Garden History Society* 1 (Fall 1991): 38–45.

Bell, Susan Groag. "Women Create Gardens in Male Landscapes: A Revisionist Approach to Eighteenth-Century English Garden History." *Feminist Studies* 16, no. 3 (1990): 471–491.

"Black Women's Club Movement." *Encyclopedia of African-American Culture and History*. https://www.encyclopedia.com/history/encyclopedias-almanacs-transcripts-and-maps/black-womens-club-movement. Accessed July 6, 2020.

Blain, Keisha N. *Until I Am Free: Fannie Lou Hamer's Enduring Message to America*. Boston: Beacon Press, 2021.

Burnett, Frances Hodgson and Tasha Tudor. *The Secret Garden*. 1st Harper Trophy ed. New York: Harper Collins, 1987.

Combahee River Collective. "'The Combahee River Collective Statement' (1977)." In *Available Means: An Anthology of Women's Rhetoric(s)*, Ritchie Joy and Ronald Kate eds., Pittsburgh: University of Pittsburgh Press, 2001: 292–300.

Crenshaw, Kimberlé. "Demarginalizing the Intersection of Race and Sex: A Black Feminist Critique of Antidiscrimination Doctrine, Feminist Theory, and Antiracist Politics [1989]." In *Feminist Legal Theory,* Boulder: Westview Press, 1991: 57–80.

Davidson, Rebecca W. *The Spirit of the Garden,* Amherst: University of Massachusetts Press, 2001.

Davis, Elizabeth Lindsay. *Lifting As They Climb*. District of Columbia: National Association of Colored Women, 1933.

Deleuze, Gilles and Félix Guattari, A Thousand Plateaus: Capitalism and Schizophrenia, Minneapolis: University of Minnesota Press, 1987; Donna Jeanne Haraway, Simians, Cyborgs, and Women: The Reinvention of Nature, New York: Routledge, 1991.

Evelyn, Charles, The Lady's Recreation: Or, The Third and Last Part of the Art of Gardening Improv'd. : Containing I. The Flower-Garden; Shewing the Best Ways of Propagating All Sorts of Flowers, Flower-Trees, and Shrubs; with Exact Directions for Their Preservation and Culture in All Particulars. II. The Most Commodions Methods of Erecting Conservations, Green-Houses, and Orangeries; with the Culture and Management of Exoticks, Fine-Greens, Ever-Greens, &c. III. The Nature of Plantations in Avenues, Walks, Wildernesses, &c. with Directions for the Raising, Pruning, and Disposing of All Lofty Vegetables. IV. Mr. John Evelyn's Kalendarium Hortense, Methodically Reduc'd: Interspers'd with Many Useful Additions. By Charles Evelyn, Esq; to Which Are Added, Some Curious Observations Concerning Variegated Greens, by the Reverend Mr. Laurence, Eighteenth Century Collections Online, London: Printed for J. Robers, near the Oxford-Arms in Warwick-Lane, 1717.

Farrand, Beatrix. *Beatrix Farrand's Plant Book for Dumbarton Oaks*. 2nd impression. Plant Book for Dumbarton Oaks. Washington, DC: Dumbarton Oaks, Trustees for Harvard University, 1993.

Farrand, Beatrix. *The Bulletins of Reef Point Gardens*. Bar Harbor, ME: Sagaponack, NY: Island Foundation; Distributed by Sagapress, 1997.

Ferrari, Carlyn Ena. *Do Not Separate Her from Her Garden: Anne Spencer's Ecopoetics*. Charlottesville: University of Virginia Press, 2022.

Ford, Charita M. "Flowering a Feminist Garden: The Writings and Poetry of Anne Spencer." *Sage* 5, no. 1 (1988): 7–14.

Frischkorn, Rebecca T., Anne Spencer, and Reuben M. Rainey. *Half My World: The Garden of Anne Spencer, a History and Guide*. Lynchburg, VA: Warwick House Pub., 2003.

Frost, H. A. and W. R. Sears. *Women in Architecture and Landscape Architecture*. Northhampton, MA: Smith College, 1928.

Gaard, Greta, "Nature, Gender, Sexuality." In *Nature and Literary Studies*, eds. P. Remien and S. Slovic, 261–279. Cambridge: Cambridge University Press, 2022.

Giddings, Paula. *When and Where I Enter: The Impact of Black Women on Race and Sex in America*. New York: Quill/W. Morrow, 1996.

Giesecke, Annette and Naomi Jacobs. *The Good Gardener? Nature, Humanity and the Garden*. London: Artifice Books on Architecture, 2015.

Glave, Dianne D. *Rooted in the Earth: Reclaiming the African American Environmental Heritage*. 1st edition. Chicago: Lawrence Hill Books, 2010.

Gordon Nembhard, Jessica. *Collective Courage: A History of African American Cooperative Economic Thought and Practice*. University Park: Pennsylvania State University Press, 2014.

Greene, J. Lee. *Time's Unfading Garden: Anne Spencer's Life and Poetry*. Baton Rouge: Louisiana State University, 1977.

Gundaker, Grey and Tynes Cowan. *Keep Your Head to the Sky: Interpreting African American Home Ground*, Democracy and Urban Landscapes. Charlottesville: University Press of Virginia, 1998.

Haraway, Donna Jeanne. *When Species Meet*. Posthumanities. Minneapolis: University of Minnesota Press, 2008.

Harris, Carmen V. "The South Carolina Home in Black and White: Race, Gender, and Power in Home Demonstration Work." *Agricultural History* 93, no. 3 (2019): 477–501.

Harris, Dianne. "Women as Gardeners," In *Encyclopedia of Gardens: History and Design*, eds. Candice A. Shoemaker and Chicago Botanic Garden. New York: Routledge, 2018: 1448.

Harris, Sheena. *Margaret Murray Washington: The Life and Times of a Career Clubwoman*. 1st edition. Knoxville: The University of Tennessee Press, 2021.

Heath, Barbara J. and Amber Bennett. "'The Little Spots Allow'd Them' The Archaeological Study of African American Yards." *Historical Archaeology* 34, no. 2 (2000): 38–55.

Henry, Marianne Morgan. "Notes: A Practical Course in Landscape Architecture." *Bulletin of the Garden Club of America* (January 1938): 122–25.

"History." *Guide to the Delphinium Garden Club of Dayton Records*. Wright State University, Special Collections and Archives, 1931–2012.

"History." National Association of Colored Women's Clubs. https://www.nacwc.com/history. Accessed August 1, 2020.

Home Demonstration Agent; AIB 38-July 1951, US Department of Agriculture, booklet; 25–27.

hooks, bell. *Belonging: A Culture of Place.* New York: Routledge, 2008.

hooks, bell. "Earthbound: On Solid Ground." In *Belonging: A Culture of Place.* New York: Routledge, 2008: 116–20.

hooks, bell. *Yearning: Race, Gender, and Cultural Politics.* Boston, MA: South End Press, 1999.

Hutcheson, Martha Brookes. *The Spirit of the Garden.* Boston: The Atlantic Monthly Press, 1923.

Hutcheson, Martha Brookes. "Wider Program for Garden Clubs." Address to the general meeting of the Garden Club of America, May 1919, New York, NY. Reprinted as "Excerpts from the Wider Program, 1919." *Bulletin* of the *Garden Club of America* (Jan. 1938): 26–27.

Jay, Martin. *Adorno.* Cambridge, MA: Harvard University Press, 1984.

Jones, Jacqueline. *Labor of Love, Labor of Sorrow: Black Women, Work, and the Family from Slavery to the Present.* New York: Basic Books, 1985.

Jones, Martha S. *Vanguard: How Black Women Broke Barriers, Won the Vote, and Insisted on Equality for All.* Basic Books, 2020.

Lee, Abra. "Profiles in Botany: Margaret James Murray Washington." *Fine Gardening,* August 15, 2022. https://www.finegardening.com/article/profiles-in-botany-margaret-james-murray-washington. Accessed January 22, 2023.

Lerner, Gerda. "Early Community Work of Black Club Women." *The Journal of Negro History* 59, no. 2 (1974): 158–160.

"Little House Gets Crop of 36 Apples." *New York Times,* July 20, 1935: 15.

Logan, Martha. *A Gardener's Kalendar,* Charleston: National Society of the Colonial Dames of America in the State of South Carolina, 1976.

McKay, George. *Radical Gardening: Politics, Idealism & Rebellion in the Garden.* London: Frances Lincoln Publishers, 2011.

McKittrick, Katherine. *Demonic Grounds: Black Women and the Cartographies of Struggle.* Minneapolis: University of Minnesota Press, 2006.

Mullins, Paul. "Gardens in the Black City: Landscaping 20th-Century African America." *Archaeology and Material Culture* (blog). July 19, 2015. https://paulmullins.wordpress.com/2015/07/19/gardens-in-the-black-city-landscaping-20th-century-african-america. Accessed August 4, 2020.

Neyland, Leedell W. and Esther Glover Fahm. *Historically Black Land-Grant Institutions and the Development of Agriculture and Home Economics, 1890–1990.* Tallahassee: Florida A & M University Foundation, 1990.

Opitz, Donald L. "'A Triumph of Brains over Brute': Women and Science at the Horticultural College, Swanley, 1890–1910." *Isis* 104, no. 1 (March 2013): 30–62. https://doi.org/10.1086/669882. Accessed August 4, 2020.

Parker, Rozsika. "Unnatural History: Women, Gardening, and Femininity" in Noël Kingsbury and Tim Richardson eds. *Vista: The Culture and Politics of Gardens.* 1st Frances Lincoln edition. London: Frances Lincoln, 2005: 87–99.

Phillips, Karen and Perry Howard. "2020 Landscape Architecture Speaker Series – Karen Phillips and Perry Howard." http://pushstudioblog.com/the-black-landscape-architects-network-2018. Accessed August 1, 2020.

Roberts, Andrea R. "The Farmers' Improvement Society and the Women's Barnyard Auxiliary of Texas: African American Community Building in the Progressive Era." *Journal of Planning History* 16, no. 3 (August 2017): 222–245.

Roberts, Andrea R., and Maia L. Butler. "Contending with the Palimpsest: Reading the Land through Black Women's Emotional Geographies." *Annals of the American Association of Geographers* 112, no. 3 (April 3, 2022): 828–837.

Scholl, Jan. "Extension Family and Consumer Sciences: Why It Was Included in the Smith-Lever Act of 1914." *Journal of Family and Consumer Sciences* 105, no. 4 (2013): 8–16.

Scott, Anne Firor. "Most Invisible of All: Black Women's Voluntary Associations." *The Journal of Southern History* 56, no. 1 (1990): 3–22.

Scott, Anne Firor. *Natural Allies: Women's Associations in American History*. Urbana: University of Illinois Press, 1991.

Shipman, Ellen Biddle. "Garden-Notebook." In *Ellen McGowan Biddle Shipman #1259,* unpaginated. Division of Rare Books and Manuscripts, Cornell University Library, Ithaca.

Spencer-Wood, Suzanne M. and Sherene Baugher. "Introduction to the Historical Archaeology of Powered Cultural Landscapes." *International Journal of Historical Archaeology* 14, no. 4 (2010): 463–74.

Sprague, Charles and William A. Sprague. "Cover Page", *Garden and Forest Magazine* 10 (1897): 192.

Thoren, Roxi. "Dreaming True." *Places Journal*, November 2018. https://doi.org/10.22269/181127. Accessed August 1, 2020.

Van Rensselaer, Mariana (Mrs. Schuyler). "Landscape Gardening—a Definition." *Garden and Forest* 1, no. 1 (1888): 2.

Van Rensselaer, Mariana (Mrs. Schuyler), *Art Out-of-Doors: Hints on Good Taste in Gardening*. New York: C. Scribner's Sons, 1893, 1925.

Waldenberger, Suzanne. "Barrio Gardens: The Arrangement of a Woman's Space." *Western Folklore* 59, no. 3 (2000): 232–245.

Walker, Alice. *In Search of Our Mothers' Gardens: Womanist Prose*. 1st edition. San Diego, CA: Harcourt Brace Jovanovich, 1983.

Waring, G. E. *Village Improvements and Farm Villages*. Boston, MA: J. R. Osgood and Company, 1877.

Washington, Booker T. *Working with the Hands: Being a Sequel to "Up from Slavery," Covering the Author's Experiences in Industrial Training at Tuskegee*. New York: Doubleday, Page & Co., 1904.

Way, Thaïsa. "Social Agendas of Early Women Landscape Architects," *Landscape Journal*, 25, no. 2 (Fall 2006): 187–204.

Way, Thaïsa. *Unbounded Practice: Women and Landscape Architecture in the Early Twentieth Century*. Democracy and Urban Landscapes. Charlottesville: University of Virginia Press, 2009.

Way, Thaïsa and Steve Callcott, "Expanding Histories/Expanding Preservation: The Wild Garden as Designed Landscape," *Preservation Education & Research Two* (2009): 53–64.

Way Thaïsa. "Longue Vue Gardens and Landscape: Ellen Biddle Shipman's Contributions." In *Longue Vue House and Gardens: The Architecture, Interiors, and Gardens of New Orleans' Most Celebrated Estate,* eds. Charles Davey and Carol McMichael Reese. New York: Skira Rizzoli, 2015, 192–221.

Way, Thaïsa, and Sahar Coston-Hardy. *Garden as Art: Beatrix Farrand at Dumbarton Oaks*. Washington, DC: Dumbarton Oaks Research Library and Collection, 2022.

White, Monica. *Freedom Farmers: Agricultural Resistance and the Black Freedom Movement*. Chapel Hill: The University of North Carolina Press, 2014.

"W. H. Manning to Frank Seiberling, Akron, Ohio, July 20, 1917." Stan Hywett Archives.

Williams, H. Hamilton. *Handbook of the Negro Garden Club of Virginia*. Hampton, VA: Hampton Institute, Department of Ornamental Horticulture and Division of Summer and Extension Study Cooperating, 1943.

Wolcott, Victoria W. *Remaking Respectability: African American Women in Interwar Detroit*. Gender & American Culture. Chapel Hill: University of North Carolina Press, 2001.

Woolf, Virginia. *A Room of One's Own*. New York: Harcourt Brace Jovanovich, 1981.

Wright, Richardson. "*House & Garden's* Own Hall of Fame." *House & Garden* (June 1933): 50.

2 Pompeian Gardens and the Archaeological Imagination

Bettina Bergmann

Woman's virtuosity lay in her containment, like the plant in the pot, limited and domesticated, sexually controlled, not spilling out into spheres in which she did not belong nor being overpowered by 'weeds' of social disorder.[1]

In the latter half of the nineteenth century, the gardens of Pompeii became a locus of desire. Painters found in the ruins the framework for an idealized past where alluring young Romans lounge in sunlit, blossoming courtyards. Unlike the earlier fashion for picturesque views of overgrown, crumbling walls, the new Pompeian genre scenes present a manicured world of daily life. There are no signs of damage or weathering; the remains, restored, are infused with an ambience of warmth, color, and intimacy. Influenced by restorations at the site itself and by the new medium of photography, the scenes are rendered with exacting precision. They are the result of a creative dialogue between archaeology and art that foreshadows today's digital reconstructions.

There is a stark contrast between the gloss of such colorful spaces and the barren ruins visitors encountered in mid-century Pompeii. Often, the inhabitant of the walled garden or secluded interior is a contemporary, upper-class woman dressed in classical costume and engaging with a plant or flowers (Figures 2.1 and 2.7). Yet no evidence survives of exclusively female spaces or activities in ancient Roman houses. The dissonance between the imagined world of antiquity in the paintings and the realities of the archaeological site reveals an imposition of European ideals of gender, nature, and domesticity. This imposition also obscures a whole other set of political and social concerns that shaped the condition of the ruins themselves, including elements as seemingly innocuous as plants.

This essay explores the open-air Pompeian scenes envisioned by two painters, the Dutch Lawrence Alma-Tadema (1836–1912) and the Bolognese Luigi Bazzani (1836–1927). Both artists spent countless hours within the excavations in the second half of the nineteenth century, and both stand out among Pompeian genre painters for their technical skills at replicating ancient materials with unprecedented archaeological accuracy. Although their styles are quite different, the two artists attempted to envision the complete buildings by combining actual with fictional elements, producing fantastic reconstructions that have shaped modern conceptions of Roman life.

DOI: 10.4324/9781003381549-3

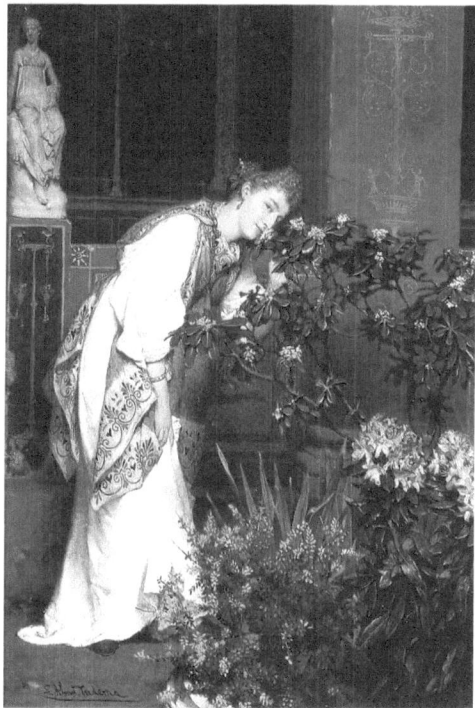

Figure 2.1 Lawrence Alma-Tadema, *In the Peristyle*, 1866. Oil on canvas, 23 x 16 in. Private Collection.

This chapter explores the artists' encounters with Pompeii, how they transformed the ruins in their paintings and the afterlives of these images. Unlike other contributions in this volume, the following is not a story of female collaboration in a garden, but rather of the construction of a man-made world where women and nature inhabit an enclosed yet liminal outdoor space.

A Culture of Reconstruction

Alma-Tadema and Bazzani came to Pompeii at a time of immense change in Italian politics, society, and archaeology. The unification of Italy in 1860 generated interest in a national past and accelerated excavations in Rome and Pompeii. In 1863, the year that Alma-Tadema first visited, the new director of the archaeological site, Giuseppe Fiorelli, began opening Pompeii to a wider public, turning it into an open-air museum and a commercial sensation. Fiorelli's aims were to preserve and display the remains as accurately as possible.[2] The site was cleared for exploration, copies of furniture and artifacts were placed in the ruins, and sketching and publishing by visitors were allowed. Fiorelli's policies intensified the exchange among scholars and artists, so that painters integrated archaeological evidence into their fictional recreations and academic publications were illustrated with pictorial scenarios of gardens and women.[3]

Thus began a new culture of reconstruction, spurred by a radical expansion in technologies of replication. As tourism swelled, so did the visual dissemination of Pompeii.[4] Photographs, postcards, and souvenir books by Neapolitan photographers, notably Giorgio Sommer, were in high demand. Their pictures of depopulated ruins, some even colorized, invited a personal viewing experience that was empty of the sightseers and workers regularly swarming the site.[5]

Fiorelli's most ambitious effort at preservation was a gigantic cork model of Pompeii, built at 1:100 scale, which was meant to replicate the entire archaeological site as it appeared by recording every faded fresco and cracked wall and allowing an overview of the excavations that could not be achieved with the panorama photographs available at the time (Figure 2.2).[6] Throughout Europe, the fashion for 3-D models inspired the creation of living reproductions to scale by wealthy patrons. The fully furnished interiors of Ludwig I's Pompejanum at Aschaffenburg (1848), the Pompeian Court of the Crystal Palace at Sydenham (1854), and the Maison Pompéienne in Paris (1855) imitated Pompeian houses while embodying modern paradigms of domesticity. The Bavarian Pompejanum, for example, was modelled on the House of the Dioscuri in Pompeii and functioned as a house-museum, presenting Roman pictorial programs on painted walls that celebrated national family values; a 1912 guidebook explains how the Roman *Hausfrau* would sit with her slaves in the women's room, planning the housework to be done. Meanwhile, the Maison Pompéienne, home to Prince Napoleon III, became a stage set for reenactments by the patron and his guests, who dressed and dined

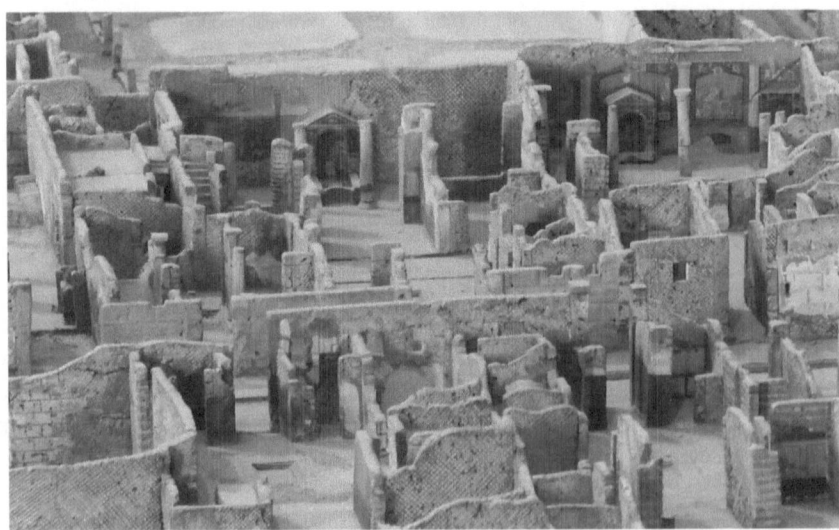

Figure 2.2 Felice Padiglione for Giuseppe Fiorelli, 1823–1896. Cork model of Pompeii. Detail: Mosaic fountains in House of the Large Fountain and House of the Small Fountain. Naples Archaeological Museum. Author photograph (Permission of Ministero per i Beni e le Attività Culturali – Soprintendenza Speciale per i Beni Archeologici di Napoli e Pompei).

like Romans.[7] The Pompeian Court was by far the best known of these immersive environments, for it was visited by millions of people and offered the template of a typical Roman house. Alma-Tadema used a photograph of the Court for his Roman spaces, which he then populated with female friends and relatives.[8]

By the end of the century, the desire for a convincing, habitable facsimile of a Roman *domus* would lead to the physical renovation of the ruined Pompeian houses themselves, along with their original furniture and even their gardens as they had supposedly appeared when Vesuvius erupted.

No doubt, the main inspiration for the human animation of visual reconstructions were Fiorelli's gypsum casts of bodies, complete with dress and jewelry, which were popularized through Sommer's photographs. The casts offered startling evidence of ill-fated Pompeians. Not surprisingly, it was the plaster forms of women that aroused special fascination and stirred fantasies about their lives. In one popular account, the cast of a female was resurrected as a "noble" who "like a good housewife [had not] forgotten her keys, after having probably locked up her store before seeking to escape;" she then fell, exposing "a limb of beautiful shape" which had formed such a perfect mould "that the cast would seem to be taken from an exquisite work of Greek art."[9] The casts solidified the role of the aestheticized female corpse as a dramatic metonym for Pompeii.

Archaeological Genre Painting

It was during this stimulating period of restitution from the 1860s on that our painters began creating scenes of Roman daily life. Despite general similarities in the two artists' pictorial recreations, their social and financial situations were quite different.

Until recently, Luigi Bazzani was little known, with many of his works unpublished, in private collections, or on the market. An instructor of perspective and set design at the Accademia di Belle Arti and later at the Accademia di San Luca in Rome, Bazzani exhibited annually in major European cities such as Vienna, Paris, Munich, and Berlin. But it was especially in Pompeii that he was prolific, producing hundreds of watercolors and drawings of buildings and frescoes for over thirty-five years, between 1880 and 1915.

In many of his views, Bazzani aspired to invoke a modern visit to the city by showing the empty ruins as they were (Figures 2.3 and 2.8). Because his precise, perspectival color renderings capture details lacking in black-and-white photographs, the chief medium of archaeological documentation, Bazzani soon came to the attention of archaeologists working at the site. By 1895, he was hired to record the excavation and rebuilding of the first grand reconstruction of a Pompeian dwelling, the House of the Vettii. He produced more than fifty drawings, sectional views, and watercolors that today are the sole evidence of lost elements of architecture and frescoes of this house.[10] Of interest here, however, is the transformation of Bazzani's exacting documentation of the remains into scenes that, like stage sets, are animated by humans and plants (Figures 2.4, 2.9 and 2.10). His expertise in scenography was especially well-suited to revivification of the ancient site.[11]

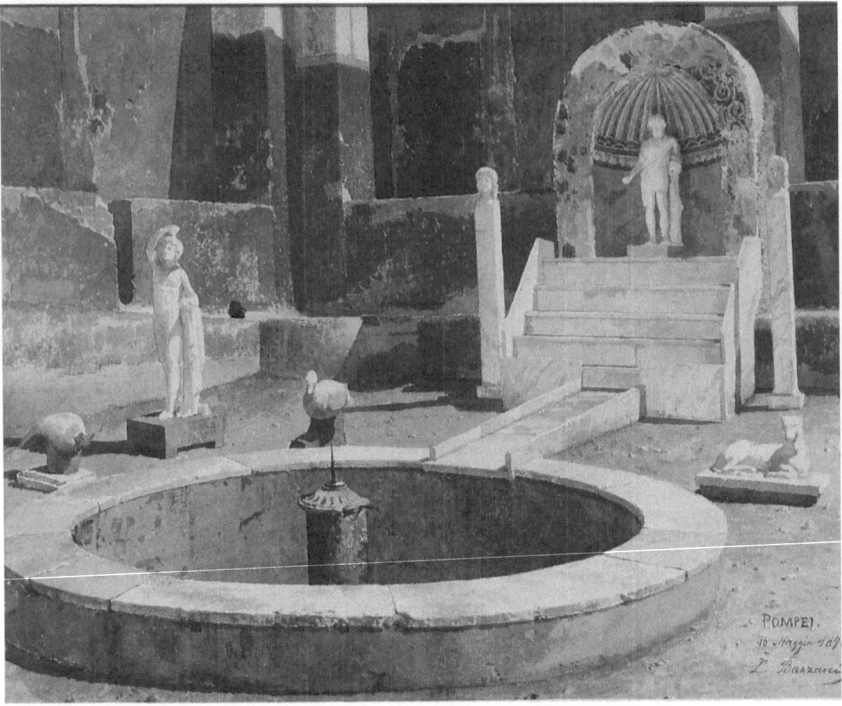

Figure 2.3 Luigi Bazzani, *View into a House with Atrium, Pompei.* Signed, inscribed, and dated *L. Bazzani, Pompei, 16 maggio 1878.* Watercolor on paper, 10 x 12.7 in. Victoria and Albert Museum, London inv. D1065-1886.

Alma-Tadema spent his entire career as an independent artist, first in Holland, until 1870, and then for the rest of his life in England. He first visited Naples and began sketching in Pompeii in 1863, just as the body casts were first being made. Shortly thereafter, in 1864, he started painting archaeological genre scenes on commission for Ernest Gambart, an influential art dealer and publisher in Antwerp who requested "pictures suited to the tastes of the bourgeoisie in England and America," namely pleasing domestic scenes.[12] From this point on, Alma-Tadema returned often to Pompeii. So much had been unearthed in the five years between his visits of 1878 and 1883 that he stayed for several weeks in 1883, working at the site every day from 9:00 am to 6:00 pm, measuring and drawing details of house after house and sketching objects. During that time, he enjoyed unusual access and boasted in a letter that Fiorelli had opened a newly excavated house just for him.[13]

However, although Alma-Tadema has been dubbed "the archaeologist of artists," he was less interested than Bazzani in reproducing whole architectural contexts. He never named a painting after a specific house but instead assembled architectural segments and dispersed artifacts to construct generic Pompeian spaces.[14] We should remember that the best objects, including extracted surfaces like wall frescoes and floor mosaics, had been removed to the Museo Borbonico in Naples, making it almost impossible to visualize accurate ensembles. Alma-Tadema depended upon visual intermediaries of 3-D replicas of statues and statuettes that he had bought in Naples and above all, upon his massive collection of 5,300 photographs, which he arranged thematically in 168 albums and used as *aide-mémoires* along with his own notes, drawings, and an extensive library of books.[15]

Alma-Tadema is often regarded as an essentially Victorian painter, but he and Bazzani were far from alone in depicting alluring women in Pompeian settings. They belonged to a large circle of cosmopolitan Neopompeian or classical-revival artists catering to a new, middle-class market. These artists' romanticized scenarios show Pompeians going about their lives with an utter lack of awareness of the city's coming destruction. In particular, images of women bathing, lounging nude, or scantily dressed, eroticize the interiors much as do the orientalizing scenes of the French "neo-Grecs."[16] In Italy, the Neapolitan painter Domenico Morelli, a friend of Alma-Tadema and one of the artists employed to replicate, in miniature, the ancient frescoes in Fiorelli's cork model, was among the first to picture living people, notably nude women, inside restored Pompeian baths, houses, and shops.[17] Unlike the depictions of women at baths and orgies, Alma-Tadema's and Bazzani's ladies are always dressed, as befits a decorous household.

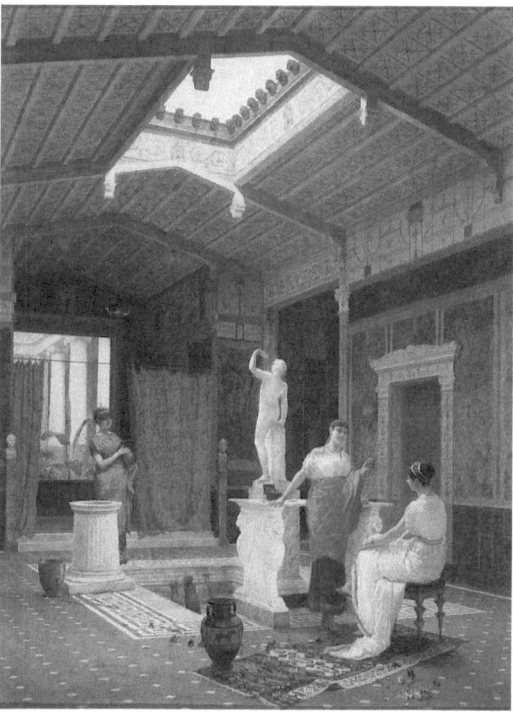

Figure 2.4 Luigi Bazzani, *A Pompeian Interior*, 1882. Oil on panel, 28 1/4 x 22 in. Signed, inscribed, and dated lower right: *Luigi Bazzani/ROMA 1882*. Dahesh Museum of Art, 1996. 24.

The shared repertory of familiar views and artifacts among archaeological genre painters—now a carved table, now a marble bench—would have resonated with prior visitors to the site, giving the viewer a sense of *déjà vu* and inviting a game of iconographic matching. Such reconstructions were, after all, called "reminiscences."[18] We can get an idea of the creative exchange, or intervisuality, among media and artists in Bazzani's *A Pompeian Interior* (1882) (Figure 2.4). Young women inhabit a Roman atrium illuminated by a skylight (*compluvium*), a water basin (*impluvium*), a marble table with winged-lion legs, and a Bacchic statue on a high base. At the time, these features could all be seen *in situ* in the House of Meleager (discovered between 1829 and 1837) and were recorded over and over again by tourists and artists.

Alma-Tadema himself used a photograph by Sommer of the interior of the House of Meleager as a model for his "Glaucus and Nydia" (1867) (Figure 2.5), a scene inspired by Edward Bulwer-Lytton's melodramatic novel *The Last Days of Pompeii* (1834), which in turn had been inspired by Karl Bryullov's enormous, much-admired painting, *The Last Day of Pompeii* (1830–33). The painting represents the scene in Book Two, Chapter 4 of Bulwer-Lytton's novel when the blind slave Nydia, hopelessly in love with the patrician Glaucus, weaves him a garland of flowers. Most of the composition is taken up with the foreground, where Glaucus reclines on a plush sofa and Nydia sits on the ground below, weaving a festoon of roses. A partial view of the room on the right shows the familiar table, marble statue on a high base, and a large oleander growing from a bronze vessel.

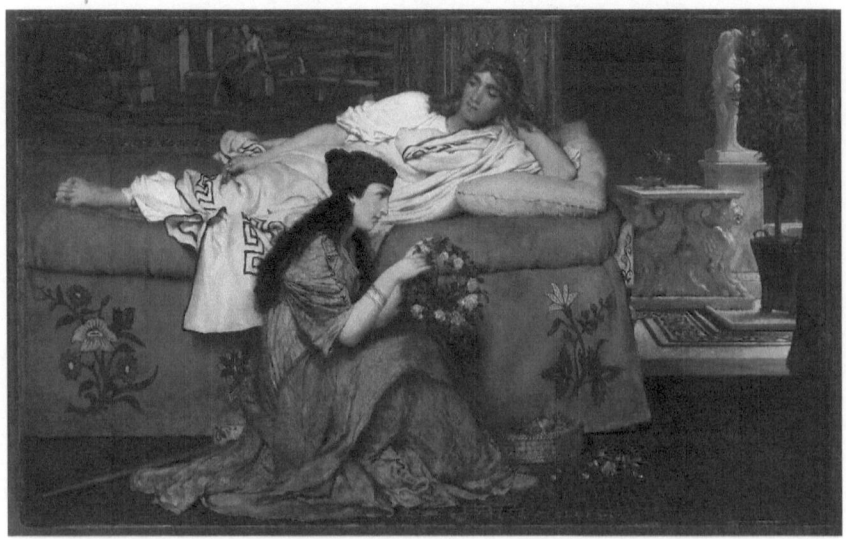

Figure 2.5 Lawrence Alma-Tadema, *Glaucus and Nydia*, 1867 reworked 1870. Oil on wood panel, 15 3/8 x 25 5/16 in. Cleveland Museum of Art. Gift of Mr. and Mrs. Noah L. Butkin 1977. 128.

Although Bazzani's and Alma-Tadema's interiors display the same artifacts, they present them from diverse angles and distances along with invented features. Compare the spacious atrium of Bazzani's *A Pompeian Interior* (Figure 2.4) with its rug, wall hangings, and view into a planted peristyle with Alma-Tadema's much shallower, densely crowded space in *Glaucus and Nydia*. Flowers are everywhere in the latter picture: in Nydia's hands, in the basket below her, in petals strewn on the floor, embroidered on the mattress cover, and blooming in the potted oleander plant. In *A Pompeian Interior* Bazzani also distributes blossoms freely; roses and flower petals lie scattered near the chatting women around the marble basin and more plants grow directly behind them in the peristyle. These paintings represent two entirely different milieux. Their originality lies in the ways that painters shaped a physical context around the now-iconic objects and animated them with human subjects.[19]

Alma-Tadema's and Bazzani's authentic scenarios received critical acclaim in public exhibitions at major European cities and appealed to consumers eager to own reproductions of ancient objects and settings. Moreover, thanks to advances in mechanical replication, their paintings were used as illustrations in books and magazines or reproduced in engravings, lithographs, etchings, and photographs that quickly sold to eager buyers. These reproductions perpetuated a set of prescribed views that the artists themselves had found in earlier photographs and postcards of Pompeii. Through repetition, an iconography of viewpoints gradually established an archetypal tourist gaze, a technology of memory.[20]

While many praised Alma-Tadema for his eye to archaeological detail and his skilled suggestion of atmosphere, art critics like John Ruskin were less enthusiastic, condemning his style of smooth, sensual surfaces, his reliance on photographs, and "packed compositions". For Roger Fry, Alma-Tadema's work was a "purely commercial art form which 'finds its chief support among the half-educated members of the lower middle-class...'."[21] Perhaps in response, or as Alma-Tadema became more confident in his work, his scenes gradually became less and less cluttered with archaeological props.

Of course, the same criticisms could apply to the style of other classical revival painters, whose photographic realism seemed mechanical to critics for whom the tangible brushstrokes of the Impressionists were revolutionary. Today, however, we can see how that realism, borne of close attention to archaeology, continues to mediate the past through film, video, and digital reconstructions.[22] It may seem puzzling that exactitude in representations of concrete things and the built environment should be enhanced — one might even say negated — by the entirely fanciful additions of fictive plants and women.

Green Spaces and Anachronisms

Much is known about the architecture and artifacts in the paintings of Alma-Tadema and Bazzani, but the living things that instill immediate sensation have received less attention.[23] What might these artists have seen of Pompeian green spaces? Although mid-century photographs show stark architectural ruins cleared of vegetation, an eager sightseer could find traces of gardens and even of some replanted areas. For example, during the Congresso Nazionale degli Scienzati in Naples in 1845, scientists were given a tour of Pompeii that included recreated gardens. Unfortunately, only one house that was visited on this tour is recorded, namely the sumptuous House of Sallust with its remains of an outdoor dining complex where a masonry couch was shaded by a pergola refreshed by a fountain and surrounded by illusionistic garden paintings; on the tour visitors saw this area replanted (though not based on actual finds) with "laurels and flowers ... roses and jasmine dispersed throughout the garden."[24]

In fact, gaining a clear picture of ancient gardens was impossible, for what a person read in ancient authors, saw depicted in frescoes, and witnessed growing in the ruins were all very different. Excavation reports occasionally mention research into vegetation, and plaster casts of tree trunks and root cavities formed a central display of the new museum Fiorelli erected on the site of Pompeii in 1873–4.[25] How such plants would have appeared in antiquity, however, was elusive. An important advance came in 1879 with the publication of *Illustrazione delle piante rappresentate nei dipinti pompeiani* by Orazio Comes, professor in Botanical Studies in the R. Scuola Superiore di Agricoltura at Portici. Comes' text, which identifies fifty plants in wall paintings, sculptures, and mosaics from Pompeii, was widely read and had great influence on the future restoration of Pompeian gardens. (Despite the title, it is unillustrated.) We should assume that Alma-Tadema and Bazzani were familiar with this work.[26]

Then there was what was growing at the site. An 1855 lithograph shows the unearthed, northern and eastern areas covered by trees. The ruins were sometimes congested with weeds, other times cleared, and sometimes landscaped anew. It was hard to tell what was typical for nineteenth-century local flora and what might have grown in antiquity. For instance, a mid-century epidemic of phylloxera led to the vineyards around Pompeii being replanted with fruit trees, which the artists would have seen during visits. At the time, no fruit trees appeared in extant Pompeian paintings, so their existence in antiquity was dismissed. Not until twentieth-century excavations revealed new physical remains and frescoes was the ubiquity of flowering fruit trees in Pompeii recognized.[27]

To add to the confusion, visitors' complaints that the archaeological site was drab and disappointing motivated archaeologists to revive the colors, textures, and scents of peristyles (a policy that continues today).[28] This impulse to reinvigorate the broken shells of buildings with a living landscape created a fascinating dilemma. The grand House of Pansa, another highlight of the sightseer's itinerary, is a case in point. When it was excavated in 1813, archaeologists found an enormous area behind the house occupying a third of the property with straight, parallel beds divided by narrow paths and, in front of the windows of a dining room, a pergola for training vines. This unusually rich evidence was described in the lavish volumes by François Mazois, *Les ruines de Pompéi* (1824–8) and in William Gell's popular *Pompeiana* (1817–19, 1832), which were in Alma-Tadema.[29] But for some reason, the garden of the House of Pansa did not catch the attention of artists. At some point in the early to mid-1800s, the backyard was transformed into a showcase for imported, exotic trees, namely orange trees, mandarin trees, lemon trees, pepper trees, magnolias, and persimmons, all of which must have delighted tourists but clearly did not reproduce original Roman landscaping.

Planting Pompeian gardens with foreign species had been routine for decades, but by the 1880s, scholars increasingly began to decry the fact that these imports from Australia, the Americas, and Asia were not only anachronistic but expressed a colonialist aesthetic. Around this time, Fiorelli and his successor Michele Ruggiero (director 1875–93) banned any further embellishment of the ruins with plants and flowers, evidently leaving previously replanted gardens as they were. However, growing knowledge about ancient flora and mounting pressure for historical accuracy reached a critical point during the last decade of the century with the first grand reconstructions of houses, as is clear from the restoration of the House of the Vettii. For the first time, in this house planting beds and paths had been discovered in the peristyle. The garden's recreation rapidly underwent two phases, which reveal the growing desire for greater authenticity. The first occurred between 1894 and 1896 when Giulio De Petra (director 1893–1900) ignored the ban by Fiorelli and Ruggiero and planted the peristyle with indigenous species, but also exotic yucca, palms, and *canna indica*.[30] Seeing this, Antonio Sogliano (Ispettore degli Scavi di Pompei off and on from 1878; director 1905–10) publicly criticized De Petra and proceeded to have the plants dug up and substituted with those mentioned by ancient authors or seen in frescoes. He relied on the book by Comes, who had identified several varieties painted on the black walls of this very peristyle, namely roses, acanthus, papyrus, hollyhocks and, previously unknown, an ingenious horticultural creation: ivy clumps trained on a stick called *metulae*. These came to life in the second recreation of the peristyle, less than two years after the first.[31]

After the disputes about the Vettii garden, at the turn of the century Sogliano fought for more faithful landscaping in Pompeian houses. He hired the academic Giuseppe Spano to undertake the first systematic use of plaster casts and identify the cavities of trees, bushes, and post-holes. Spano returned to houses that had been excavated decades earlier, found new garden beds, and replanted spaces according to the finds. He was aided by Nicola Roncicchi, the topographer and gardener who worked on Fiorelli's cork model of Pompeii. Roncicchi made a significant discovery in the House of the Centenary (1902), namely roots of oleanders and small flowers spaced at a regular distance from each other, a horticultural technique now known to have been common in Roman gardens. He placed new plants in terracotta pots in the earth, just as (he learned) ancient gardeners had done. More root cavities were excavated in the House of the Golden Cupids between 1903 and 1905, which led to even more accurate replantings, and by 1906 the peristyles of the best-preserved houses had all been recreated according to the latest evidence.[32] This was the birth of garden archaeology.

Alma-Tadema and Bazzani must have been aware of these developments and seen the changes in vegetation growing at the site. Indeed, Bazzani witnessed the embattled restoration of the House of the Vettii firsthand when he recorded the planting beds in sketches and watercolors. Yet knowledge of what was authentic to ancient gardens was, and still is, fluid. In the flurry to replace foreign species with more faithful ones, it was not recognized that from the first century BCE, the Romans themselves were busily importing and grafting exotic plants, which they cultivated in hothouses and showed off in peristyles and back gardens.[33] Ironically, the anachronistic plantings of mid-nineteenth century Pompeii were much in the spirit of Roman horticulture.

Did Alma-Tadema's and Bazzani's garden spaces reflect the growing knowledge about plants, or did they evoke Pompeian taste for exotica?

Alma-Tadema's Flowers

It is noteworthy that plants, and especially flowers, are often the main feature of Alma-Tadema's paintings. One of his most popular works, *Hearty Welcome – Corner of a Roman Garden* (before 1878) (Figure 2.6), depicts a spacious, flowering courtyard where a Roman family returns home. The columns and roof tiles betray the artist's close inspection of peristyles; the covered arbor he likely modeled on the one reconstructed in the celebrated Villa of Diomedes at Pompeii; and the fountain, statuette of Eros, and lararium were all documented in the artist's photograph collection. The plants, on the other hand, raise intriguing questions.

Such profuse growth is unlikely in an ancient courtyard. Compositionally, the large sunflowers growing against the garden wall and the swathe of bright red poppies in the foreground harmonize with the yellow and red ochre of walls and columns. But where did Alma-Tadema see them? Poppies grew wild both in ancient and nineteenth-century Italy, usually in dense masses within a walled garden. Sunflowers, though, were first grown as a crop by indigenous tribes in North America and not taken to Europe until 1500, when they became widespread as an ornamental. Alma-Tadema may have seen the majestic yellow flower during his visits to gardens in Rome and Campania. But an ancient Pompeian never would have, and for this apparent mistake he was much criticized. Was this an oversight? As we shall see, the painter enjoyed mixing and matching extant and fantastic features in his imaginary Roman gardens.[34]

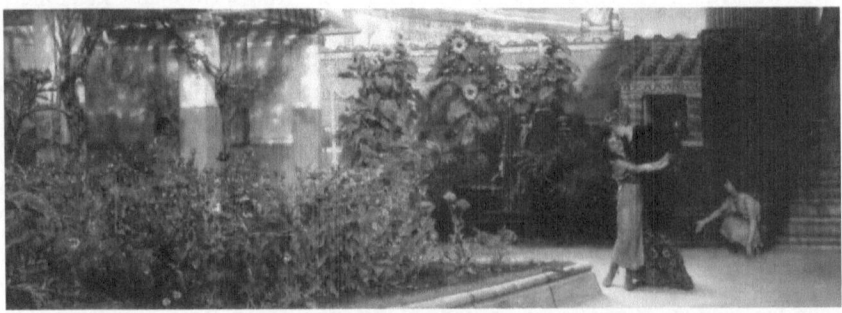

Figure 2.6 Lawrence Alma-Tadema, *Hearty Welcome – Corner of a Roman Garden*, before 1878. Oil on canvas, 11.81 x 35.83 in. The Ashmolean Museum of Art and Archaeology WA1955. 60.

What is more, the family in *Hearty Welcome* is not anonymous: the father, obscured by shadow descending the stairs at the right, is in fact Alma-Tadema himself; the girl below him who kneels to greet a dog, is his daughter Laurense, and the central group his younger daughter Anna embracing her stepmother Laura.[35] Indeed, Alma-Tadema regularly used living models for his Roman figures. In *In the Peristyle* (1866) (Figure 2.1), his first wife, the red-haired Marie-Pauline Gressin-Dumoulin de Boisgirard, wears a resplendent white dress and glances back at us as she bows to inhale the bouquet of a blooming white rhododendron. Our eye moves from the marble statue of a goddess to Marie-Pauline to the variety of flowering plants, all set against bold red and black painted walls.[36] As is typical for Alma-Tadema's classical scenarios, woman and plant invoke the sensuality of an olfactory experience.

The highly acclaimed *An Oleander* (1882) (Figure 2.7), painted some four years after *Hearty Welcome*, suggests that Alma-Tadema may have read Comes' book and noted that *Nerium Oleander L.* was the most common plant depicted in Pompeii. An enormous, solitary oleander rises from a large metal vessel in an interior courtyard.[37] To the right, a young woman in a transparent, green chiton with loose sleeves sits on the edge of a marble basin filled with exotic shells and raises an

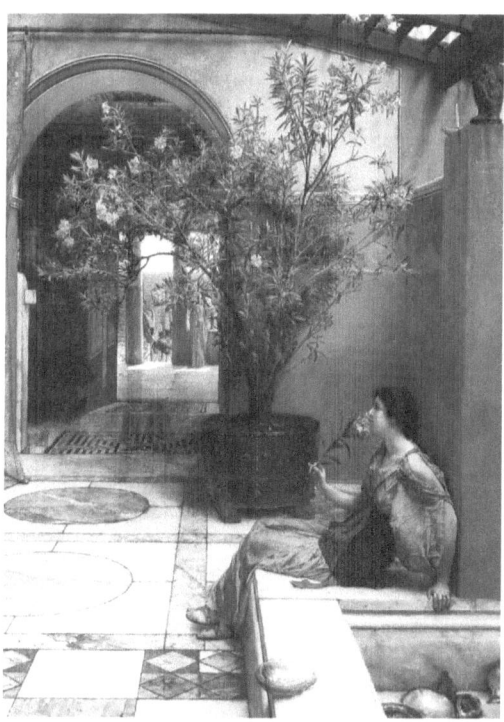

Figure 2.7 Lawrence Alma-Tadema. *An Oleander*, 1882. Signed and inscribed: *L Alma Tadema op. CCXLV* (upper right). Oil on panel, 35.9 x 25.7 in. Private Collection.

oleander flower to her nose to enjoy its scent. The rich red walls behind her contrast with the distant view of blue sea where a party departs by boat towards an illuminated horizon.

The artist's effective cropping carries the eye from the woman and pink blossoms against the red wall to the oleander's long branches that reach across a corridor; we move from a room saturated in warm, red, and rosy shades to the cool caerulean sea.[38] The setting is unlike that of ancient interiors, whose walls and ceilings were never solid colors but luminescent frescoes embellished with lively ornaments and figures. The marble-encrusted floor, too, would have been rare in a Pompeian townhouse, as would such a view, both of which were far more likely enjoyed from an imperial palace or elite coastal villa. As was his custom, Alma-Tadema fashioned wealthy Roman residences of luxury where moments of leisure and retrospection occur naturally.

The oleander in Alma-Tadema's picture plays the main protagonist, illuminated from above, taller and more vibrant than the young woman seated at the lower right. While it is true that oleanders were ubiquitous in Pompeii, they would not have been planted individually in a pot as here nor would a solitary shrub have dominated an interior space. Instead, oleanders grew in groups with other plants in open, spacious plots. As a matter of fact, this was not just any oleander, but a beloved specimen that Alma-Tadema himself nurtured in his London home for a decade and which suddenly burst into bloom in 1881, a year before he finished the painting. His friend Georg Ebers wrote:

> Once Tadema was fascinated for weeks by a large oleander which he had taken with him from Brussels to London, where for the first time it covered itself with blossoms. He had no peace until he had succeeded in fixing upon canvas the wealth and delicacy of its luxuriant rose-hued flowers ....[39]

Alma-Tadema, then, painted plants from life just as he did portraits of friends and family. In his conservatory he cultivated the oleander along with poppies, sunflowers, roses, cala lilies, jonquils, forget-me-nots, hawthorn, cherry and apple trees, and ivy. Some he had imported — roses from the French Riviera and Africa — recalling the habit of Roman aristocrats who ordered roses from southern Italy and Egypt for their extravagant dinner parties. When in the winter months between 1890 and 1894 he was at work on *Spring*, a steady stream of deliveries ensured a continually fresh supply of budding plants, and for botanical accuracy, he consulted his many photographs.

Often in Alma-Tadema's compositions flowers form the true subject and the entire scene becomes a kind of still life, a habit that some have attributed to his Dutch training.[40] Flowers feature in over half of the more than 340 classicizing scenes that he created from 1865 and may have been inspired by the Roman gardens he and his family visited in 1863.[41] In 1878 Alma-Tadema and his wife Laura again spent three months travelling on the Continent; he wrote of Italian plants: "Oranges and lemons, olives and spring flowers, brown sunkissed mankind young and old, graceful and strong, sometimes very beautiful. Fine art and antiquity."[42] Thus in his own day, Alma-Tadema was labeled a flower painter.

> No other artist has ever made so much use of flowers to beautify his pictures as Alma-Tadema. They frequently aid him in his difficulties of color and composition. A picture which will not come right is often settled by a mass of splendid bloom from his garden or conservatory. In this respect he allowed himself some liberty of anachronism … introducing the latest variety of purple clematis or rose azalea into the gardens and palaces of ancient Rome.[43]

This last comment no doubt refers to the blazing purple azalea in *Unconscious Rivals* (1893); the "liberty of anachronism" allowed him to introduce a glamorous exotic into a barren architectural environment.[44] Historical accuracy, we are reminded, was not his goal.

Alma-Tadema's flowers are routinely regarded as purely compositional elements providing color and balance and have no historical value, but some do add a narrative dimension.[45] Consider his historical paintings. The bold red poppies dominating the foreground of *Tarquinius Superbus* (1867) are specifically mentioned by Livy as portentous, and *The Roses of Heliogabalus* (1888) offers a vivid condemnation of imperial greed and decadence. Clearly Alma-Tadema drew on both ancient and modern connotations of flowers. In some cases, he was inspired by ancient, male, bucolic poets' ideas about love and nature, specifically by the *Anacreontea*, a collection of some sixty Greek poems dating from the first century BCE to the sixth century CE. Roses play a special role. Like other British classical-subject painters, he consulted the urbane Roman poets — Catullus, Tibullus, Propertius, Ovid, Horace — for whom roses signal love. For the Victorian audience knowledgeable about the Language of Flowers, scenes where men are approaching women with roses would have evoked the rituals of courtship.[46]

The flowers of Alma-Tadema, in short, were polyvalent, transportable from context to context, like the living plants themselves. Some, it has been suggested, may relate to the death and resurrection of Pompeii. Poppies drop their petals as soon as they are picked. The anemone, like the poppy, grew wild around Pompeii; its name derives from the Greek for windflower, signifying fragility and transience, for the delicate flowers are blown open by the wind, which then blows away the dead petals. And although roses, Alma-Tadema's most popular flower, resonate with ancient poets and Victorian flower symbolism, they can also invoke mortality, as in *The Last Roses* (1872), where a young woman places dying autumn roses on a marble altar.[47]

Alma-Tadema's Roman green spaces, adorned with gleaming marble and floral profusion and inhabited by contemporary women, were his personal blend of here and then. In this way, the artist aligned with the current Victorian concept of a mirror of history, whereby the classics formed the "furniture of the mind" for the educated person. This was his personal world. He decorated his two studio-houses to resemble Roman villas, and by placing his wife and daughters within an ancient environment, he established a nearness to what is lost and distant. In that collapse, women inhabit a domestic paradise with no hint of labor or industry, like birds in a gilded cage.

Not all viewers have been positive about this temporal collapse.

What makes his pictures ludicrous is their unbelievability. His Roman maids and matrons typically behave like proper English ladies. They tend to fill their sitting rooms with 19th-century flowers (hybrid roses, rhododendrons) from their 19th-century gardens ... exceptionally tidy, bourgeois .... These modest, well-scrubbed women are so happily engrossed in their proper English lives they scarcely seem to notice they've been transported whole to Pompeii or Capri.[48]

Bazzani's Courtyards

If flowers, color, and sensual surfaces make up the essence of Alma-Tadema's Pompeian gardens, for Bazzani it was architecture, marble furniture, and water features.[49] One can get an idea of such features in green spaces that were visible

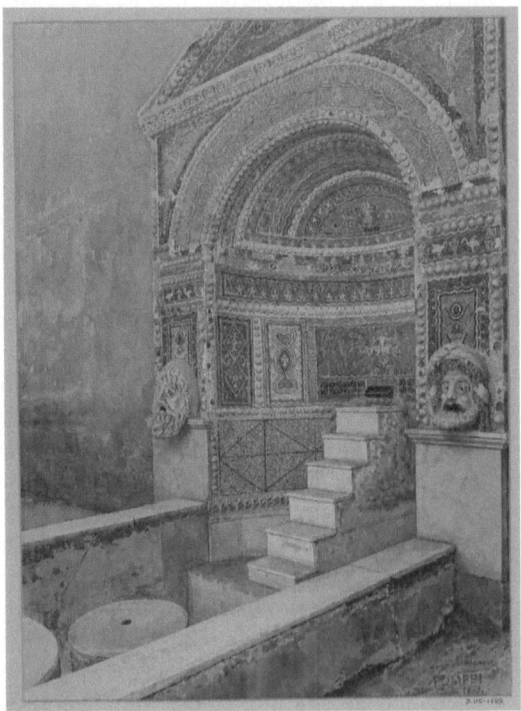

Figure 2.8 Luigi Bazzani. *Mosaic Fountain with Tragic Masks*. House of the Large Fountain (VI.8.22) Pompeii, 1889. Inscribed *Luigi Bazzani POMPEI 1889*. Watercolor on paper, 8.2 x 10.9 in. Victoria and Albert Museum. Inventory number D.115–1889.

to both artists from the cork model displaying the excavations from 1861 through 1879, which unearthed several peristyles with illusionistic murals, statuary, pools, and wall fountains (Figure 2.2).

Although much at the site and in the model had faded by the latter part of the nineteenth century, still vivid were the beautifully preserved mosaic fountains in the House of the Large Fountain (found in 1826), the House of the Small Fountain (found in 1827), and the House of the Grand Duke of Tuscany (excavated in 1833 and 1845).[50] Bazzani was especially drawn to these scenographic installations of small edifices covered with glistening cut glass, which in antiquity formed the focal point of houses for passersby on the street. Beginning with detailed watercolors of the remains, Bazzani went on to visualize completed spaces in oil, producing nothing short of archaeological reconstructions. For one example, his watercolor of the House of the Grand Duke of Tuscany records in detail the fragmentary garden mural behind the fountain with its steps for cascading water, three basins, and a marble table in front. In a later painting, he has rebuilt the upper walls, paved the ground with a black-and-white mosaic, restored the illusionistic mural with plants and swinging garlands on the walls, pumped water into the fountain, and enlivened the space with a young woman who stands, gazing down at the water in the basins.[51] A similar process of fastidious recording, this time a watercolor of the mosaic fountain in the

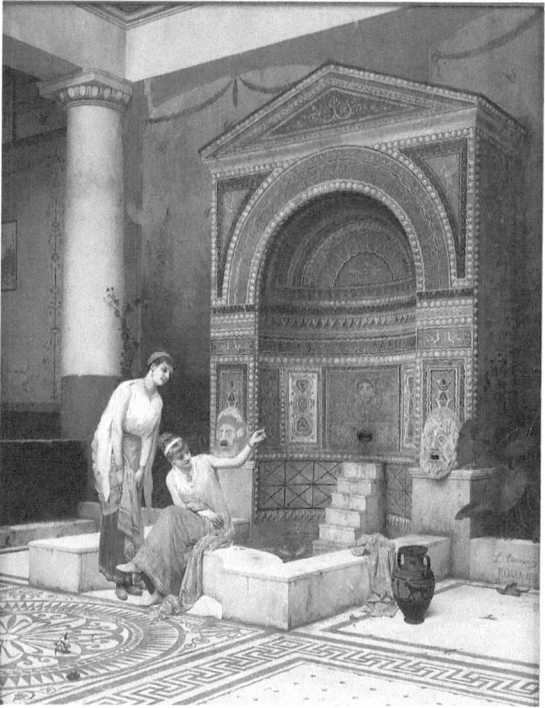

Figure 2.9 Luigi Bazzani. *Women Feeding Fish in a Pompeian Atrium*, 1879. Oil on panel, 23 × 29 in. Signed, inscribed, and dated.

House of the Large Fountain (Figure 2.3), led to the oil painting *Women Feeding Fish in a Pompeian Atrium* (1879) (Figure 2.9). Working in his Rome studio, he placed two women at the edge of the pool. And to the right of the fountain, he put *colocasia* (elephant ears), a plant native to Asia, Australia, and India.

Bazzani became particularly fascinated with the House of Marcus Lucretius (IX.3.5.24). Excavated in 1846–7, this dwelling became the most visited, photographed, and sketched house in the second half of the nineteenth century, especially sought out for the theatrical vista from its entrance, which opened onto an elevated, stage-like planted area with a large basin, an ornate fountain, and eighteen statues.[52] Bazzani spent much time on the garden. He documented its current state in several sketches and watercolors, embellishing it to various degrees, a process that offers an illuminating insight into the artist's working methods. His watercolor *View into a House with Atrium, Pompeii* (signed and dated *L. Bazzani, Pompei, 16 maggio 1878*) is the first in a series of studies on the house (Figure 2.3). It is interesting, considering his love of scenography, that he did not adopt the popular view from the street but brings the viewer up onto the garden stage. The focal point is the mosaic *aedicula* with its step fountain carrying water down into a large round basin. Framed by marble herms, a statue of Silenus stands in the niche, and statuettes of animals and male deities surround the basin on the ground.[53]

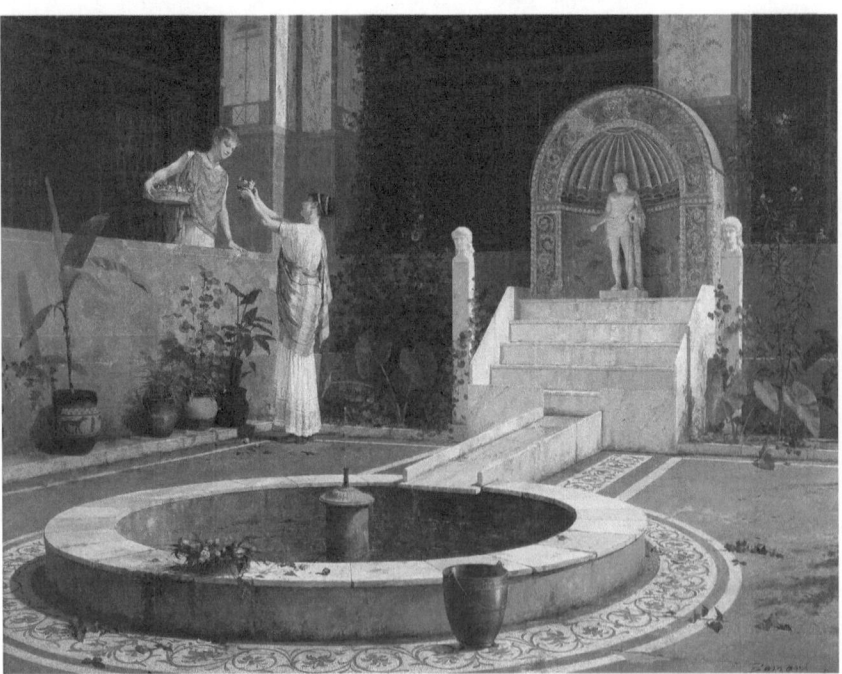

Figure 2.10 Luigi Bazzani. *Picking Flowers from the Courtyard*, before 1927(1878–9?). Signed *Bazzani* and inscribed *Roma* (lower right). Oil on panel, 17.5 x 22.9 in. Private Collection.

These studies formed the basis for the oil painting *Picking Flowers from the Courtyard* (before 1927) (Figure 2.10).[54] Two women have stopped their activities to interact; one stands in a portico above, a basket of flowers cradled in her arm, while the other in the courtyard below extends a handful of loose flowers up to her. For some reason, Bazzani omits the statuettes of animals and deities that he had so carefully recorded and either introduces elements that were found elsewhere or invents others. At the time, the paths and low hedges around the basin were obscured by weeds, so he moved the black-and-white floor mosaic from a nearby room (*tablinum*) opening onto this garden. The resulting courtyard with its hard floors and rebuilt walls leaves little room for living elements, but a variety of carefully rendered plants — a climbing mandevilla vine with red blossoms and elephant ears (both exotic), and an oleander with pink flowers — rise from beds on either side of the fountain, while green plants in clay pots line up along the wall. More features signal the passing of time: water shoots from the jet inside the basin, a basket of freshly picked flowers sits on its rim, and a recurring attribute of Bazzani's Pompeian spaces, petals and branches are scattered on the marble basin and the ground.

Both Alma-Tadema's and Bazzani's recreated gardens display plants in clay and metal containers. Two terracotta pots were found to the left of the fountain niche in the House of Marcus Lucretius, but ornamental vessels were rare in Roman gardens, and certainly there were none like the Greek vases in Bazzani's courtyards, which the artist might have seen on view in the Naples museum. Almost all the clay pots found in excavations are pierced with holes and embedded in the ground for training vines or transplanting species.[55] Because at the time little was known about planting practices, contemporary Italian gardens offered ready examples for our painters.

One ornamental with large, pointed leaves, the banana tree, appears in so many of Bazzani's reconstructed courtyards that it seems to be his signature plant (Figures 2.4, 2.10). In *Picking Flowers from the Courtyard*, it grows in an extremely large terracotta pot painted with black figures (Figure 2.10).[56] In *Pompeian Interior* (1875), a young woman leans forward in a pose echoing the marble statue of Flora behind her and picks a white flower, possibly a rose; to either side of her, the same tall bush with pointed leaves grows in two huge clay vessels.[57] In another painting with a similar title, *A Pompeian Interior* (1882) (Figure 2.4), a banana plant forms the focal point of the view into the peristyle. In fact, these all are based upon a single specimen that Bazzani rendered in two pencil drawings, one of banana leaves and the other of a banana tree in a pot.[58]

Bananas are a tropical crop. According to the Greek naturalist Theophrastus and the Roman naturalist Pliny the Elder, Alexander the Great brought it to the west from his Asiatic campaign in 331–323 BCE, and the emperor Augustus is said to have promoted cultivation of the African banana (*musa ornata*) (Suetonius *August.* 59, 81). It is unclear whether it could be grown in Italy or ever reached Pompeii. Bazzani's bananas and elephant ears add a frisson of exoticism.

The question remains: what accounts for the perpetual association of attractive young women and plants in idealized Roman gardens? The question opens out in multiple directions.

Plotting Pompeian Gardens

Unlike the other chapters in this book, this chapter does not describe an arena of female collaboration, but stands as a counterpoint to it. It describes an arena of female isolation, where elegantly attired Roman women stand or sit in open but enclosed courtyards and atria, areas connected to the home yet permeable through columns, walls, doors, and windows. Accompanied by plants and flowers, the women appear focused on private reverie or intimate conversations.[59]

This vision of domesticity arose from the encounter of nineteenth-century painters with Pompeian excavations. Their settings are not just credible because they feature actual artifacts, but because the medium of oil painting could achieve what black-and-white photographs, or even a visit to the site, could not. The colors and textures of pigments evoke the sensory stimuli of warm sunlight, cool marble, and fragrant blossoms, and material ruins come to life through the introduction of the ephemeral — of passing clouds, flowing water, living beings.[60] The verisimilitude of the painted scenes made them look natural, with the result that they became iconic images of classical antiquity in the later realistic media of cinema and digital reconstruction.

Bazzani and Alma-Tadema were just two of many artists who propagated the pictorial fantasy of passive women in a rarefied domestic realm. Scenes of Roman daily life were wildly popular among the new, middle-class market in Europe and the United States, not only among men but women as well. By the 1880s and 1890s, thousands of visitors were traveling to see genre-paintings by Alma-Tadema, Bazzani, and their contemporaries in the art galleries and exhibitions of major cities.[61] Attractive young women in ancient gardens continued as a pictorial subject into the early 1900s, notably in the works of William Godward (1861–1922) and John William Waterhouse (1849–1917).[62]

Alma-Tadema never sought to register what a nineteenth-century tourist saw in Pompeii. In his corpus, classical gardens are just one theme among many, while Bazzani devoted his career to recording as closely as possible the current state of the ruins. Unlike Alma-Tadema's imaginary settings, his recreations retain the architectural framework of buildings, to which he adds elements from other houses, figures, and plants, in some cases exotic ones. These hardscapes lack the sensuality infused by Alma-Tadema's graceful humans and dazzling flowers or the intimacy achieved through his artful cropping. No female relatives dominate the scene; rather, pairs or triads of anonymous Roman women remain relatively proportionate to their setting. Both artists, however, inject time and immediacy into a static setting with cultivated plants and domesticated women, turning the lens onto blooming and breathing living things.

Their gardens are revealing. Alma-Tadema and Bazzani were working at a time of growing knowledge and heated controversy about planted spaces, as archaeologists increasingly strove to recreate the authentic landscaping of Pompeian peristyles. These physical site reconstructions were based upon the actual remains emerging from the soil together with the study of ancient texts and frescoes. In our painters' gardens, elements of architecture, fountains, marble tables, and statues often correspond to the excavated remains, but plants such as, anemones, poppies, azaleas, elephant ears, and bananas are out of place, imported or planted separately in pots and courtyards. These anachronistic additions resulted in liminal spaces where past and present, local and foreign seem to fuse.

Ancient gardens, we now believe, were primarily green, with formal beds of low shrubs and clipped box bushes lining paths or a garden wall. Flowers played a small role. Occasionally a large fruit tree might offer shade in a peristyle.[63] These green spaces inside Pompeian houses often were oriented to the view of visitors and passersby, for they advertised the amenities of space, light, water, and foliage. A viewer passing an open front door would rarely, if ever, have glimpsed a solitary, well-dressed woman admiring a bloom. Indeed, one has difficulty imagining a Pompeian peristyle without the presence of men or slaves.

Despite the anachronisms, Alma-Tadema and Bazzani still continue to influence the ways that we imagine ancient gardens.[64] Their enduring sway is due in large part to the ubiquitous reproductions of their paintings, which have been acquired or circulated in books and magazines. Yet the afterlives of the two artists follow divergent trajectories. It is Bazzani's watercolors of archaeological ruins, and not his holistic recreations of Pompeian spaces, that are in demand today. In truth, his meticulous recording of ruins, rendered in linear perspective and naturalistic color, present far more convincing semblances of the areas exposed by archaeology than many recent digital reconstructions, for their luminous pastels capture the tones and textures of faded frescoes on exposed, sunlit walls in ways that photographs and computers cannot.

Alma-Tadema's paintings have found a wider, more popular dissemination. His pictorial fantasies were emulated by other painters even during his lifetime. They were staged as tableaux in live performances and live on in cinema.[65] Ridley Scott chose Alma-Tadema's works as models for his settings and costumes in *Gladiator* (2001), intentionally shunning the available archaeological evidence.[66] As his production designer, Arthur Max, explained:

> Ridley and I decided we wouldn't do the classical, scholastic Rome, which could be represented by researching and staying totally faithful to the museum archives and the factual concepts of Roman scholars. We decided we were more impressed by the romantic vision of Rome by painters such as Alma-Tadema .... We tried to emulate the accessories, pageantry, opulence and scale in his paintings.... Instead of dead archaeology we wanted antiquity to be alive and breathing.[67]

"Alive and breathing" antiquity was achieved in nineteenth-century genre paintings through female figures and their flowers. Although the association between women and plants may have been a literary trope among ancient poets,[68] it was not so much there as in contemporary fiction that the artists would have encountered female bodies animating Pompeian ruins. Specifically, Edward Bulwer-Lytton's *The Last Days of Pompeii* (1834), Théophile Gautier's short story "Arria Marcella: Souvenir of Pompeii"(1852), and later, Wilhelm Jensen's novel *Gradiva* (1902), envisioned the female form as enabling a veritable rebirth of antiquity.[69] As Catharine Edwards states, one "strategy for coming to terms with the mysterious strangeness of the ancient past was to gender it as feminine."[70] And writing of travelers' literature in the late eighteenth and early nineteenth centuries, Chloe Chard says that

> Feminised ruins also supply an especially useful means of transmuting historical time into personal time .... By defining an ancient fragment or ruin as a site of a ghostly female presence, the traveller ... makes the vestige of antiquity more easily transportable into a private domain of emotional intimacy ... deploying a woman as a means of shifting ancient history into an intimate, private world ...."[71] In reconstructions of the classical past, the female presence idealizes a gender binary and captures the vexed experience of not seeing while being seen.[72]

It may seem paradoxical that men painted such gendered scenarios at a time when women were gaining agency as writers, painters, and, notably, as gardeners too (although in Britain, schoolgirls were taught how to nurture plants in order to acquire the feminine virtues of patience and caring).[73] Alma-Tadema's first and second wives and daughters were successful artists; they may have served as his models, but in life they were far from confined to the interior and quite active within the public realm. In Alma-Tadema's portrait of his fifteen-year-old daughter, *Miss Anna Alma-Tadema* (1883), the girl faces directly out at the viewer and raises a vase of large white carnations. Anything but domestic and demure, Anna was a suffragist whose own paintings emulated her father's portraits, interiors, and flowers and which she showed at national exhibitions along with her stepmother. Her sister Laurense, a poet, novelist, critic, playwright, and short story author, was an ardent activist and world traveler. Less is known about the wives and daughters of the Bazzani family, but as with the Alma-Tadema household, there seems to have been lively familial collaboration. Several of the men were practicing architects or artists, painting portraits and, like Luigi, scenes of Pompeii. Luigi's wife, Elena Fracassini-Serafini, was the sister of the eminent painter Cesare Fracassini, who worked together with Bazzani on public buildings in Rome. In Italy, as elsewhere, this was a time when creative women were gaining authority and influence.[74]

Considering the growing activities of European women, it is significant that Alma-Tadema's and Bazzani's pictures show no females writing or painting, no garden gloves or tools. Instead, the women are as manicured and eye-catching as the statues and plants beside them. Barred from the public sphere and thus from present society, they do not work as artists, writers, or gardeners. They are as contained and domesticated as a plant in a pot, and containment precludes agency and collaboration.

The act of reconstruction is never innocent. Through a process of selection, interpretation, and cultural appropriation it creates a cultural narrative that shapes our understanding of the past.[75] The second half of the nineteenth century saw a fruitful exchange between art and archaeology in which ruins served to authenticate painters' (and buyers') visions of a seductive antiquity of beauty and pleasure. Their recreations of outdoor planted spaces, in particular, reveal the slippage between accuracy and fantasy.

Such scenes must have offered refuge from the anxiety and social upheaval caused by industrialization. It is noteworthy that in France in the very same years the subject of women relaxing in flower gardens became a prevalent theme among Impressionists. But in contrast, those gardens were actual gardens enjoyed as oases within the modern-day, urbanized landscape of France, and not as an alternative, classicizing reality.[76]

The desire of Fiorelli to preserve and complete the ruins of Pompeii continues as the ancient garden moves back and forth between drawing, photograph, watercolor, oil painting, and digital and actual physical recreations. In today's immersive museum installations, 3D tours, and virtual reality, one might encounter Roman women moving through spectral spaces as holograms or projections. Such augmented reality experiences can seem more real than the sites themselves, perhaps in a similar way to the nineteenth-century realistic oil paintings of classical gardens. For those visiting Pompeii in person, the replanting of gardens within the ruins turns assumptions into living, material facts. All are simulations that distort our understanding of what is authentic. The scenarios of Alma-Tadema and Bazzani remind us of how the present will always reshape the past, and also of how that recreated past paints the present.

Notes

1 Davidoff and Hall, *Family Fortunes*, 191–2.
2 For a contemporary account, see Monnier, *The Wonders of Pompeii*.
3 Among the many examples, Niccolini and Niccolini, *Le case ed i monumenti di Pompei.* On Fiorelli's influence on genre paintings, see Nwokobia, "Picturing Pompeii", 184–5.
4 Kovacs, "Pompeii and Its Material Reproductions," 37–8; Gardner Coates, "On the Cutting Edge"; Urry, *The Tourist Gaze*, 3; Crary, *Techniques of the Observer*.
5 On Sommer: Gardner Coates and Seydl, *Antiquity Recovered*, 221–4; Miraglia, "La fotografia, Pompei e l'Antico"; De Carolis "Pompei in posa nella fotografia dell'Ottocento"; Ascione, "Il 'souvenir' di Pompei"; Ascione, "Tra vedutismo e fotografia."
6 Work on the model proceeded from 1861 to 1879, with a pause at the end of the century, resuming in 1908 with the landscaper Nicola Roncicchi: Sampaolo, "La realizzazione del plastico di Pompei"; Kockel, "I modelli di Pompei," 267–75.
7 Bergmann, "A Tale of Two Sites,"183–215; Hales, "Re-casting Antiquity," 115; Hales, "Living with Arria Marcella," 217–44; Blix, *French Romanticism and the Cultural Politics of Archaeology,* 209–16.
8 The Pompeian House in the Crystal Palace was inspired by descriptions and illustrations of the House of the Tragic Poet in Gell and Gandy, *Pompeiana* and of the House of Glaucus in Bulwer-Lytton, *The Last Days of Pompeii*. On Alma-Tadema and the Pompeian House, see Nichols, *Greece and Rome at the Crystal Palace,* 111–13.
9 "Pompeii," *Quarterly Review*, 115, no. 230 (1864): 312–48, 322. Dwyer, "Science or Morbid Curiosity?" 171–88; Dwyer, *Pompeii's Living Statues*; De Caro, "Excavation and Conservation at Pompeii," 16–18. Moormann, *Pompeii's Ashes*; Nwokobia, "Picturing Pompeii", 67–76; on the ruins of Pompeii as appealing to a "picturesque melancholy" of the Victorian imagination: Hales, "Cities of the Dead".

10 Scagliarini, Coralini, and Helg, *Davvero!*; Flamini and Prisco, "La casa dei Vettii ai tempi della scoperta"; Cassanelli, "Pompeii in Nineteenth-Century Painting," 46; for many images of Bazzani's paintings, see https://www.artrenewal.org/artists/luigi-bazzani/1639.

11 Helg, "Luigi Bazzani, pittore e scenografo,"161–72; "Luigi Bazzani, un artista nella Pompei di fine '800"; "Tra vedutismo e documentazione"; "Dall'osservazione della realtà all'invenzione"; "Antichità da immaginare, antichità da documentare"; Rambaldi, "Echi pompeiani ed ercolanesi."

12 On Gambart: Verhoogt, *Art in Reproduction*, 42–329. When Gambart retired in 1871, the affiliation continued with Léon Henri Lefèvre.

13 Alma-Tadema returned to Pompeii in 1878, 1881, 1883–4, and 1896: Barrow, *Use of Classical Past*, 28–41; Nwokobia *Picturing Pompeii*, 91. For color images of many paintings: https://www.artrenewal.org/artists/lawrence-alma-tadema/8.

14 A nickname given by his friend in Antwerp, the German Egyptologist Georg Ebers in *The Nation*, 1888. On Alma-Tadema's working method, see Prettejohn, "Antiquity Fragmented and Reconstructed," 33–44; Moser, *Painting Antiquity*, 31–8, 142; Rovira-Guardiola, "Archaeology in Alma-Tadema's Painting."

15 Pohlmann, "Alma-Tadema and Photography," 111–16; Moser, *Painting Antiquity*, 381–4; Miraglia, "La fotografia, Pompei e l'Antico," 31–55; Nwokobia, "Picturing Pompeii," 116, 142.

16 Alma-Tadema met Jean-Léon Gérôme in Paris in 1864. On Gérôme and the *Néo-Grecs* or *les Pompéistes:* Becker and Prettejohn, *Sir Lawrence Alma-Tadema*, 69–76. Moser, *Painting Antiquity*, 311–14; Nwokobia, "Picturing Pompeii," 23–8.

17 The international circle of genre painters and Alma-Tadema's relationship to Italian artists are discussed by De Caro and Querci, *Alma Tadema e la nostalgia dell'antico*. The popularity of the "School of Posillipo" culminated at the Esposizione Nazionale d'Arte di Napoli in 1877; Ascione, "Pompei e il mondo classico," 87–9; Figurelli, "Italian Classical-Revival Painters," 136–52; Irollo, "Artisti, opere e mercato fra Napoli e Londra," 86–97; Blix, *French Romanticism,* 204–16; Besnard, *Le retour à l'Antique*; Berardi, "I 'pittori archeologi' nella Roma postunitaria"; Couelle, "Désirs d'Antique" and "Alma-Tadema ou les couleurs del'Antiquité"; Prettejohn, "Recreating Rome in Victorian Painting," 35–42.

18 Prettejohn, "Lawrence Alma-Tadema and the Modern City of Ancient Rome,"120, 127: "Even individual artifacts evince a peculiar doubleness; they simultaneously display comforting signs of survival and disconcerting traces of loss."

19 On Glaucus and Nydia, see Prettejohn, *At Home in Antiquity*, 21, Fig. 5; Gardner Coates et al., *The Last Days of Pompeii,* 202, Nr. 70.

20 On the iconography of the tourist gaze and the mutual influence of photos and paintings upon each other, see Hartnett, "Excavation Photographs and the Imagining of Pompeii's Streets," 257–62; Urry, *The Tourist Gaze.*

21 On the negative reception: Verhoogt *Art in Reproduction*, 2–329; Moser, *Painting Antiquity*, 332–7; Swanson, *The Biography and Catalogue Raisonné*, 95. Ruskin, *The Art of England*, 68–71; Fry, "The Case of the Late Sir Lawrence Alma Tadema", 149. In the *Burlington Magazine* (Feb. 2013): 286, Arthur Clutton-Brock complained that Alma-Tadema had tried to make the viewer believe that he had witnessed the moments represented in his pictures by filling them with archaeologically accurate details.

22 Piccoli, *Visualizing Antiquity before the Digital Age.*

23 On painters' representations of gardens, see Ascione, "Il giardino pompeiano nelle vedute neoclassiche e romantiche," 69–90.

24 Casalena, "The Congresses of Italian Scientists"; Ciarallo, *Gli spazi verdi dell'antica Pompei*, 175, 210; Stefani and Borgongino, "I Giardini Pompeiana tra Realtà e Finzione," 29.

25 Dwyer, *Pompeii's Living Statues*, 106–8.

26 Comes, "Illustrazione delle piante rappresentate nei dipinti pompeiani" followed the first academic publication on Roman gardens by Schouw (1851). On the history of

plant research in Pompeii: Francisscen, "A Century of Scientific Research on Plants"; Jashemski, *Natural History,* 80–4.

27 For speculation about what Alma-Tadema saw of trees and vines in 1863, see Nwokobia, "Picturing Pompeii," 164–7. For identifications of the modern flora in Pompeii, see Ciarallo and De Carolis, *Around the Walls of Pompeii.*

28 On visitors' reactions: Beard, "Taste and the Antique: Visiting Pompeii in the Nineteenth Century."

29 Gell's *Pompeiana* (1817–19) was the first English account of remains at Pompeii. His imaginary view through the atrium to peristyle garden with fountain (Plate 36) offered readers a novel virtual reconstruction, which influenced Neopompeian buildings such as the Crystal Palace.

30 Sogliano, *La casa dei Vettii,* 233–388.

31 Stefani and Borgongino, "I Giardini Pompeiana tra Realtà e Finzione," 29; Ciarallo *Gli spazi verdi dell'antica Pompei,* 211.

32 Jacono, *Osservazioni su i viridarii pompeiani,* gives a quick tour of the newly replanted gardens. On the replanting by Roncicchi: Zevi, *Pompei 1748–1980,* 18. On the archaeobotany and patterns of trade and cultivation in Pompeii: Jashemski et al. *Gardens of the Roman Empire,* 469–71; Nwokobia, *Picturing Pompeii,*140, 180.

33 Macaulay-Lewis, "Imported Exotica."

34 Alma-Tadema as a flower painter: Zimmern, *Sir L. Alma-Tadema, Royal Academician*; Johnston and Lovett, *Empires Restored, Elysium Revisted*; Nwokobia, *Picturing Pompeii*, 90. He designed his gardens at Grove End Road like his paintings. In turn, his paintings influenced the eminent landscape architect, Harold Peto (1854–1933): Helmreich, "The Gardens at Grove End Road," 165.

35 Swanson, *The Biography and Catalogue Raisonné,* nn. 236 and 202. On the mistake of sunflowers: n. 181; Nwokobia "Saying It with Flowers," 56–61; Nwokobia, *Picturing Pompeii,* 223–30.

36 Also called *In the Conservatory*: Swanson, *The Biography and Catalogue Raisonné* nn. 75, 139.

37 Barrow, *Lawrence Alma-Tadema,* 92, no. 85 with previous literature. The painting was commissioned by Leon Henri Lefevre in 1882 and first exhibited at the Royal Academy of Arts. Presently, it is owned by a Spanish-Mexican businessman, Juan Antonio Perez Simon.

38 Zimmern, *Sir Lawrence Alma Tadema,* 23; Johnston and Lovett, *Empires Restored, Elysium Revisited,* 184–5, n. 30; Swanson, *The biography and Catalogue Raisonné,* nn. 281, 218–19.

39 Ebers, *Lorenz Alma Tadema,* 56.

40 For example, in *Flowers* (1868), Museum of Fine Arts, Boston, a woman leans into the center of the painting to inhale the fragrance of potted plants of carnations, roses, forgetme-nots, rhododendrons: Johnston and Lovett, *Empires Restored, Elysium Revisited,* 57 n. 12; Swanson, *The Biography and Catalogue Raisonné,*150.

41 On the flowers in *Spring* (1895) see Lippincott, *Lawrence Alma-Tadema: Spring,* 42–4; Swanson, *The Biography and Catalogue Raisonné,* 49.

42 Letter to F. G. Stevens, 18 April 1878, F. G. Stevens Archive, Bodleian Library, Oxford.

43 Monkhouse, "Laurens Alma-Tadema, R.A.," 672.

44 On *Unconscious Rivals,* see Prettejohn, "Recreating Rome in Victorian Painting," 67.

45 Lippincott, *Lawrence Alma-Tadema: Spring,* 42–4 sees the flowers selected for their color or symbolism as "unreliable historic signposts."

46 Casteras, *Images of Victorian Womanhood*; Landow, "Victorianized Romans"; Barrow, "The Scent of Roses." The roses of Campania were especially famous and sacred to Pompeii's patron goddess Venus: Jashemski, *Natural History of Pompeii,* 158–60. Pompeii: http://pompeiisites.org/en/projects/the-ancient-rose-of-pompeii.

47 Zimmern, *Sir L. Alma-Tadema, Royal Academician,* 15; 21 on the poppies in *Tarquin* and "*Hearty Welcome.* Red poppies were common in antiquity: Verg. *Georgics*

1.207–14; Ciarallo, *Gli spazi verdi*, 142. Nwokobia, Saying It with Flowers, suggests that many flowers relate to death and rebirth in Pompeii.

48 Review of the exhibition "Empires Restored, Elysium Revisited: The Art of Sir Lawrence Alma-Tadema" at the Stirling and Francine Clark Art Institute of Williamstown, MA, by Paul Richard, *Washington Post,* February 25, 1992. https://www.washingtonpost.com/ archive/lifestyle/1992/02/25/art/2e3b6a86-f366-4ebf-b687-063f15b76540.

49 On Bazzani's gardens, see Helg, "Luigi Bazzani, pittore e scenografo," 166–8.

50 House of the Grand Duke of Tuscany (VII.4.56): Jashemski, *The Gardens of Pompeii* Vol. 1, 41; Vol. 2, 180–1, 362; House of the Little Fountain (VI.8.23): Jashemski, *The Gardens of Pompeii* Vol. 2, 136, figures 149–52; House of the Great Fountain (VI.8.22/1): Jashemski, *The Gardens of Pompeii* Vol. 2, 135.

51 Luigi Bazzani, "Mosaic fountain with three pools in Pompeii" (before 1927) watercolor: https://collections.vam.ac.uk/item/O406356/drawing/Luigi Bazzani, "Pompeii atrium"(before 1927): https://commons.wikimedia.org/wiki/File:Pompeii_atrium_by_ Luigi_Bazzani,_before_1927.jpg

52 Niccolini and Niccolini, "Casa di Marco Lucrezio"; Jashemski, *The Gardens of Pompeii* Vol. 2, 231, 366; Hales and Leander Touati, *Returns to Pompeii,* 253–71. Alma-Tadema had three photographs of this garden, which became the background for *Catullus at Lesbia's* (1865).

53 Three subsequent sketches in pencil on paper render the same features from varying viewpoints and distances. In one, Bazzani restores the missing upper walls. In others, he offers a close-up of the fountain and more detailed objects from a side view: Helg, "Luigi Bazzani, pittore e scenografo," 166; pencil on paper sketches: Benucci, "Fontana pompeiana e foglie di pianta esotica (banana)," 1.1–1.3; Benucci, "Viridarium della case di Marco Lucrezio."

54 Helg, "Dall'osservazione della realtà," 28–9 describes Bazzani's method of combining elements from different houses.

55 Jashemski, "Vasa Fictilia," 371–91.

56 On model sketches for this painting, see Leardi, *I Bazzani a Pompei*, 1.6. Bazzani's many studies of plants have yet to be examined.

57 Luigi Bazzani, "Pompeian Interior" 1875, Oil on panel, 17 ⅜ x 12 ½ in, St. Johnsbury Athenaneum, Vermont, Gift of Horace Fairbanks.

58 Benucci, "Fontana pompeiana e foglie di pianta esotica (banana)." The banana plant in a pot drawn on tissue paper appears in another oil painting, Cortile di una casa a Pompei (Private Collection, Rome).

59 Bailes, "Women, Gardens, and Solitude". On women in nineteenth-century art, see Pollock, "Modernity and the Spaces of Femininity," 70–127.

60 Savani, "Sensing the Past"; Hamilakis, *Archaeology and the Senses.*

61 Goldhill, *Victorian Culture and Classical Antiquity,* 23.

62 Nwokobia, *Picturing Pompeii*, 198–203.

63 Jashemski, *The Gardens of Pompeii* Vol. 1, 53–4.

64 De Caro, "Excavation and Conservation at Pompeii,"18–19 states that Neopompeian painters influenced reconstructions of the House of Silver Wedding (excavated 1891– 1908), the House of Centenary (1879, 1881, 1902), the House of the Vettii (1894–5), and the House of the Golden Cupids (1905).

65 Verhoogt, "Reproducing Alma-Tadema." Amedeo Maiuri, the principal director of Pompeii in the twentieth century, commissioned Bazzani to illustrate his 1928 survey of the site: Maiuri, *Pompei, con 14 acquerelli originali di Luigi Bazzani.*

66 Alma-Tadema created stage designs for historical plays: Becker and Prettejohn, *Sir Lawrence Alma-Tadema,* 258–9. Blom states that through cinema, Alma-Tadema's image of antiquity became our own: "The Second Life of Alma-Tadema"; Barrow, *The Use of the Classical Art,* 222. Films include Cecile B. DeMille's *Ten Commandments"* (1956) and *Cleopatra* (1934), which drew from *Spring. The Flower Market* Opus LXII (1868) inspired a tableau in James Pain's pyrodrama (1879–1914) and the film Gladiator (2000): Swanson, *The Biography,* 1990, 150, n. 10. Prettejohn, *At Home in Antiquity,*

171 points out that Tadema continues to inspire filmmakers around the world, including South Koreans.

67 Quoted in Blom, "The Second Life of Alma-Tadema,"196–7.

68 In contrast to the sweet roses and maidens of Greek poetry, Latin authors place irreputable women in parks and gardens, where insubordination finds free rein: Pagán, *Horticulture and the Roman Shaping of Nature*.

69 Hales, "Living with Arria Marcella," 217–44; on the female body as a conduit to resurrection, see Betzer, "Afterimage of the Eruption"; Betzer, "Archaeology Meets Fantasy," 118–35.

70 Edwards, *Roman Presences,* 16.

71 Chard, "The Road to Ruin", 128.

72 Pollock, "Modernity and the Spaces of Femininity," notes how artists use spaces to differentiate women, specifically "dining rooms, drawing rooms, bedrooms, balconies/verandas, private gardens." Kestner, *Mythology and Misogyny,* 283, states that Alma-Tadema's "patriarchal gaze" and "classicism served reactionary ends."

73 Page and Smith, *Women, Literature, and the Domesticated Landscape*; Fanucci, *Women Gardeners,* Page "Gardening for Women," 51–67 with previous bibliography; Davidoff and Hall, *Family Fortunes,* 191–2; 373 on women's identification with flowers and the training of young girls; Alexander, "The Garden as Occasional Domestic Space"; Bliston, "Queens of the Garden." In 1864, Ruskin delivered two lectures published as "On Queens'Gardens" (later *Of Sesames and Liles,* London, 1904), which promoted the Victorian idea of separate spheres and a woman's essential character belonging within the confines of home.

74 Sambuco, *Italian Women Writers*; Frau and Gragnani, *Sottoboschi letterari* 2011.

75 On the "rhetoric of archaeology" see Blix, *French Romanticism and the Cultural Politics of Archaeology*. For the use of the past to visualize the present: Lowenthal, *The Past is a Foreign Country*; Nora, "'Between Memory and History."

76 Herbert, *Impressionism,* 177–93.

References

Alexander, Catherine. "The Garden as Occasional Domestic Space." *Signs* 27.3 (Spring 2002): 857–71.

Ascione, Gina Carla. "Tra vedutismo e fotografia: la rappresentazione di Pompei nella seconda metà dell'Ottocento." In *Fotografi a Pompei nell'800. Dalle collezioni del Museo Alinari.* Catalogo mostra Pompei Scavi, Casina dell'Aquila, 5 dicembre 1990–6 aprile 1991, edited by Baldassare Conticello, Claudio de Polo Saibanti, 21–9. Florence: Alinari, 1990.

Ascione, Gina Carla. "Il giardino pompeiano nelle vedute neoclassiche e romantiche." In *Domus – Viridaria – Horti picti,* Le Mostre, Soprintendenza archeologica di Pompei, 69–90. Pompeii: Bibliopolis, 1992.

Ascione, Gina Carla. "Il 'souvenir' di Pompei. Dalle immagini neoclassiche alla diffusione nell'epoca della riproducibilità tecnica." *Rivista di Studi Pompeiani* 12/13 (2001–2002): 35–51.

Ascione, Gina Carla. "Pompei e il mondo classico nella produzione pittorica napoletana tra 'accademia' a 'storia'." In *Storie di un Eruzione. Storie da un'eruzione. Pompei, Ercolano, Oplontis,* edited by Pietro Giovanni Guzzo, Marisa Mastroroberto, and Antonio d'Ambrosio, 84–93. Naples: Mondadori Electa, 2003.

Bailes, Melissa. "Women, Gardens, and Solitude in Eighteenth-Century Britain." *The Eighteenth Century* 57.4 (Winter 2016): 537–41.

Barrow, Rosemary. "The Scent of Roses: Alma-Tadema and the Other Side of Rome." *Bulletin of the Institute of Classical Studies* 42 (1997–98): 183–202.

Barrow, Rosemary. *Lawrence Alma-Tadema*. London: Phaidon Press, 2001.

Barrow, Rosemary. *The Use of Classical Art and Literature by Victorian Painters, 1860–1912*. Lewiston, Queenston, and Lampeter: The Edwin Mellen Press, 2007.

Beard, Mary. "Taste and the Antique: Visiting Pompeii in the Nineteenth Century." *Studies in the History of Art*, Vol. 79, Symposium Papers LVI: *Rediscovering the Ancient World on the Bay of Naples, 1710–1890*, edited by Carol Mattusch, 205–28. Washington, DC: National Gallery of Art, 2013.

Becker, Edwin and Elizabeth Prettejohn, eds. *Sir Lawrence Alma-Tadema*. New York: Rizzoli, 1997.

Benucci, Michele. "Viridarium della case di Marco Lucrezio Stabia con ipotesi ricostruttiva," In Geraldine Leardi, ed. *I Bazzani a Pompei: I Disegni e acquarelli nell'archivio di stato di Terni*. Terni: Archivio di Stato, 2016a, 1.1–1.3.

Benucci, Michele. "Fontana pompeiana e foglie di pianta esotica (banana),", In *I Bazzani a Pompei: I Disegni e acquarelli nell'archivio di stato di Terni*, edited by Geraldine Leardi, 1.9–1.10. Terni: Archivio di Stato, 2016b.

Berardi, G. "I 'pittori archeologi' nella Roma postunitaria e il signor Goupil." In *Alma Tadema e la nostalgia dell'antico*, Catalogo della mostra Napoli 2007–8, edited by Stefano De Caro and E. Querci, 98–109. Milan: Electa, 2007.

Bergmann, Bettina. "A Tale of Two Sites: Ludwig I's Pompejanum and the House of the Dioscuri in Pompeii." In *Returns to Pompeii. Interior Space and Decoration Documented and Revived. 18th–20th Century,* edited by Anne-Marie Leander and Shelley Hales, 183–215. Stockholm: Swedish Institute in Rome, 2016.

Besnard, Tiphaine. *Le retour à l'Antique: entre véracité archéologique & fantasme: 1840–1910. Art et histoire de l'art*. n.p. 2014.

Betzer, Sarah. "Afterimage of the Eruption: An Archaeology Chassériau's *Tepidarium* (1853)." *Art History* (2010). https://onlinelibrary.wiley.com/doi/full/10.1111/j.1467-8365.2010.00765.x.

Betzer, Sarah. "Archaeology Meets Fantasy: Chassériau's Pompeii in Nineteenth-Century Paris." In *Pompeii in the Public Imagination from Its Rediscovery to Today,* edited by Shelley Hales and Joanna Paul, 118–35. Oxford: Oxford University Press, 2011.

Bliston, Sarah. "Queens of the Garden. Victorian Women Gardeners and the Rise of the Gardening Advice Text." *Victorian Literature and Culture*. 36.1 (2008): 1–19.

Blix, Göran. *French Romanticism and the Cultural Politics of Archaeology*. Philadelphia: University of Pennsylvania Press, 2009.

Blom, Ivo. "The Second Life of Alma-Tadema." In *Lawrence Alma-Tadema: At Home in Antiquity*, edited by Elizabeth Prettejohn and Peter Trippi, 187–99. Munich: Prestel, 2016.

Casalena, Maria Pia. "The Congresses of Italian Scientists between Europe and the Risorgimento (1839–75)." *Journal of Modern Italian Studies* 12.2 (2007): 153–88.

Cassanelli, Roberto. "Pompeii in Nineteenth-Century Painting." In *Houses and Monuments of Pompeii: The Works of Fausto and Felice Niccolini,* edited by Roberto Cassanelli, Pier Luigi Ciapparelli, Enrico Colle, Massimilano David, 40–7. Los Angeles: The J. Paul Getty Museum, 2002.

Casteras, Susan P. *Images of Victorian Womanhood in English Art.* Rutherford, NJ: Fairleigh Dickinson University Press; London: Associate University Presses, 1987.

Chard, Chloe. "The Road to Ruin: Memory, Ghosts, Moonlight and Weeds." In *Roman Presences: Receptions of Rome in European Culture, 1789–1945*, edited by Catherine Edwards, 125–39. Cambridge: Cambridge University Press, 1999.

Ciarallo, Annamaria and Chiara Giordano. *Gli spazi verdi dell'antica Pompei*. Rome: Aracne, 2012.

Ciarallo, Annamaria and Ernesto De Carolis, eds. *Around the Walls of Pompeii. The Ancient City in Its Natural Environment*, Soprintendenza Archeologica di Pompei, Milan: Electa, 1998.

Clutton-Brock, Arthur. "Alma Tadema." Burlington Magazine 22. 118 (1913) 285–7.

Comes, Orazio. *Illustrazione delle piante rappresentate nei dipinti pompeiani.* Naples: F. Giannini, 1879.

Couëlle, C. "Alma-Tadema ou les couleurs del'Antiquité." *Travaux & documents,* Université de La Réunion, Faculté des lettres et des sciences humaines (2007): 135–74.

Couëlle, C. "Désirs d'Antique ou comment rêver le passé grécoromain dans la peinture européenne de la seconde moitié du XIXe† siècle." *Anabases. Traditions et réceptions de l'Antiquité* 11 (2010): 1–30.

Crary, Jonathan. *Techniques of the Observer: On Vision and Modernity in the Nineteenth Century,* Cambridge, MA: MIT Press, 1990.

Davidoff, Leonore and Catherine Hall. *Family Fortunes: Men and Women of the English Middle Class, 1780–1850* (1987). Chicago: University of Chicago Press, 2002.

De Caro, Stefano. "Excavation and Conservation at Pompeii: A Conflicted History." *The Journal of Fasti Online: Archaeological Conservation Series* (2015): 1–31. www.fastionline.org/docs/FOLDER-con-2015-3.pdf.

De Caro, Stefano and E. Querci, eds. *Alma Tadema e la nostalgia dell'antico.* Catalogo della mostra Napoli 2007–2008, Milan: Mondadori Electa, 2007.

De Carolis, Ernesto. "Pompei in posa nella fotografia dell'Ottocento." In *Pompei e l'Europa. 1748–1943.* Catalogo della mostra, edited by Massimo Osanna, M. T Caracciolo and Luigi Gallo, 277–85, 294–300. Milan: Mondadori Electa, 2015.

Dwyer, Eugene. "Science or Morbid Curiosity? The Casts of Giuseppe Fiorelli and the Last Days of Romantic Pompeii." In *Antiquity Recovered: The Legacy of Pompeii and Herculaneum,* edited by Victoria Coates and Jon Seydl, 171–88. Los Angeles: J. Paul Getty Museum, 2007.

Dwyer, Eugene. *Pompeii's Living Statues: Ancient Roman Lives Stolen from Death.* Ann Arbor: University of Michigan Press, 2013.

Ebers, George. "The Archaeologist of Artists." *The Nation* (New York), 16 September 1886, 237–8.

Ebers, George, *Lorenz Alma Tadema; His Life and Works*, from the German by Mary J. Safford, New York: Leopold Classic Library, 1886.

Edwards, Catherine, ed. *Roman Presences: Receptions of Rome in European Culture, 1798–1945.* Cambridge: Cambridge University Press, 1999.

Fanucci, Pola. *Women Gardeners: Stivali, penne e pennelli di giardiniere appassionate.* ETS Publishing, 2016.

Figurelli, Luna. "Italian Classical-Revival Painters and the "Southern Question'." In *Pompeii in the Public Imagination from Its Rediscovery to Today,* edited by Shelley Hales and Joanna Paul, 136–52. Oxford: Oxford University Press, 2011.

Flamini, Maria Grazia and Gabriella Prisco, "La casa dei Vettii ai tempi della scoperta: Luigi Bazzani, un testimone d'eccezione." In *Davvero! La Pompei di fine '800 nella pittura di Luigi Bazzani,* edited by Daniella Scagliarini, Antonella Coralini, Riccardo Helg, 40–7. Bologna: Fondazione del Monte di Bologna e Ravenna, 2013.

Francisscen, Frans P. M. "A Century of Scientific Research on Plants in Roman Mural Paintings (1879–1979)." *Rivista di Studi Pompeiani* 1 (1987): 111–22.

Frau, Ombretta and Gragnani, Cristina, eds. *Sottoboschi letterari: sei case studies fra Otto e Novecento: Mara Antelling, Emma Boghen Conigliani, Evelyn, Anna Franchi, Jolanda, Flavia Steno.* Florence: Firenze University Press, 2011.

Fry, Roger. "The Case of the Late Sir Lawrence Alma Tadema, O.M.," *The Nation*, 18 January 1913, reprinted in *A Roger Fry Reader*, edited by Christopher Reed, 147–8. Chicago, IL: University of Chicago Press, 1996.

Gardner Coates, Victoria C. "On the Cutting Edge: Pompeii and New Technology." In *The Last Days of Pompeii: Decadence, Apocalypse, Resurrection*, edited by Victoria C. Gardner Coates, Ken Lapatin, and Jon Seydl, 44–51. Los Angeles: The J. Paul Getty Museum, 2012.

Gardner Coates, Victoria C., Ken Lapatin, and Jon Seydl, eds. *The Last Days of Pompeii: Decadence, Apocalypse, Resurrection*, Los Angeles: The J. Paul Getty Museum, 2012.

Gardner Coates, Victoria C. and Jon L. Seydl, eds. *Antiquity Recovered: The Legacy of Pompeii and Herculaneum*, Los Angeles: The J. Paul Getty Museum, 2007.

Gell, William. *Pompeiana: The Topography, Edifices and Ornaments of Pompeii: The Result of Excavations since 1819*, 2 vols, London: Rodwell and Martin, 1832.

Gell, William and John Gandy. *Pompeiana: The Topography, Edifices and Ornaments of Pompeii*, 2 vols, London: Chatto and Windus, 1817–19. (New ed. 1824. Further edition by Gell alone incorporating the results of latest excavations. London 1832 and 1852.)

Goldhill, Simon. *Victorian Culture and Classical Antiquity: Art, Opera, Fiction, and the Proclamation of Modernity*, Princeton, NJ: Princeton University Press, 2011.

Hales, Shelley. "Cities of the Dead." In *Pompeii in the Public Imagination. From Its Rediscovery to Today*, edited by Shelley Hales and Joanna Paul, 153–70. Oxford: Oxford University Press, 2011.

Hales, Shelley. "Re-casting Antiquity: Pompeii and the Crystal Palace." *Arion* 14.1 (2006): 99–133.

Hales, Shelley. "Living with Arria Marcella: Novel Interiors in the Maison Pompéienne." In *Returns to Pompeii: Interior Space and Decoration Documented and Revived 18th–20th Century*. Skrifter utgivna av Svenska Institutet i Rom. Serie in 4°, 62, edited by Shelley Hales and Anne-Marie Leander Touati, 217–44. Stockholm: Svenska institutet i Rom, 2016.

Hales, Shelley and Joanna Paul, eds. *Pompeii in the Public Imagination. From Its Rediscovery to Today*, Oxford: Oxford University Press, 2011.

Hales, Shelley and Anne-Marie Leander Touati, eds. *Returns to Pompeii: Interior Space and Decoration Documented and Revived 18th–20th Century*. Skrifter utgivna av Svenska Institutet i Rom. Serie in 4°, 62. Stockholm: Svenska institutet i Rom, 2016.

Hamilakis, Y. *Archaeology and the Senses: Human Experience, Memory, and Affect*. Cambridge: Cambridge University Press, 2014.

Hartnett, Jeremy. "Excavation Photographs and the Imagining of Pompeii's Streets: Vittorio Spinazzola and the Via dell'Abbondanza." In *Pompeii in the Public Imagination from Its Rediscovery to Today*, edited by Shelley Hales and Joanna Paul, 246–69. Oxford: Oxford University Press, 2011.

Helg, Riccardo. "Luigi Bazzani, pittore e scenografo: un bolognese a Pompei." *Il Carrobbio* 32 (2006): 159–75.

Helg, Riccardo. "Tra vedutismo e documentazione: gli acquerelli pompeiani di Luigi Bazzani." In *Atti del X Congresso Internazionale dell'AIPMA (Association Internationale pour la Peinture Murale Antique) Napoli, 17–21 settembre 2007*, edited by Irene Bragantini, 889–92. *AION AnnAStorAnt* 18, 2010.

Helg, Riccardo. "Luigi Bazzani, un artista nella Pompei di fine '800." In *Davvero! La Pompei di fine '800 nella pittura di Luigi Bazzani*, edited by Daniella Scagliarini, Antonella Coralini, and Riccardo Helg, 19–27. Bologna: Fondazione del Monte, 2013a.

Helg, Riccardo. "Dall'osservazione della realtà all'invenzione: le opere 'neopompeiane'." In: *Davvero! La Pompei di fine '800 nella pittura di Luigi Bazzani*, edited by Daniella Scagliarini, Antonella Coralini, and Riccardo Helg, 27–31. Bologna: Fondazione del Monte, 2013b.

Helg, Riccardo. "Antichità da immaginare, antichità da documentare. Il ruolo dell'archeologia nell'attività professionale di Luigi Bazzani." In *I Bazzani a Pompei: disegni e acquerelli nell'Archivio di Stato di Terni*, edited by Geraldine Leardi, 78–93. Terni: Archivio di Stato, 2016.

Helmreich, Anne. "The Gardens at Grove End Road." In *Lawrence Alma-Tadema: At Home in Antiquity*, edited by Elizabeth Prettejohn and Peter Trippi, 165–6. Munich: Prestel, 2016.

Herbert, Robert L. *Impressionism: art, leisure, and Parisian society*. New Haven, CT: Yale University Press, 1988.

Irollo, Alba. "Artisti, opere e mercato fra Napoli e Londra: Appunti su Alma-Tadema, Amendola e Morelli." In *Alma-Tadema e la nostalgia dell'antico*, exhibition catalogue (Naples, Museo Archeologico Nazionale, October 19, 2007–March 31, 2008), edited by Stefano De Caro and Eugenia Querci, 86–97. Rome: Electa, 2007.

Jacono, Luigi. *Osservazioni su i viridarii pompeiani*. Torre Annunziata: Officina graf. E. Letizia, 1910.

Jashemski, Wilhelmina. *The Gardens of Pompeii: Herculaneum and the Villas Destroyed by Vesuvius*. Vol. 1. New Rochelle, NY: Aristide d Caratzas, 1979.

Jashemski, Wilhelmina. "Vasa Fictilia: Ollae Perforatae." In *The Two Worlds of the Poet: New Perspectives on Vergil*, edited by Robert M. Wilhelm and Howard Jones, 371–91. Detroit: Wayne State University Press, 1992.

Jashemski, Wilhelmina. *The Gardens of Pompeii: Herculaneum and the Villas Destroyed by Vesuvius: Appendices*. Vol. 2. New Rochelle, NY: Aristide d Caratzas, 1993.

Jashemski, Wilhelmina. *The Natural History of Pompeii*. Cambridge: Cambridge University Press, 2002.

Jashemski, Wilhelmina, Kathryn Gleason, Kim J. Hartswick, and Amina-Aïcha Malek, eds. *Gardens of the Roman Empire*, Cambridge and New York, NY: Cambridge University Press, 2018.

Johnston, William R. and Jennifer G. Lovett, eds. *Empires Restored, Elysium Revisted: The Art of Sir Lawrence Alma-Tadema*, Williamstown, MA: Sterling and Francine Clark Art Institute, 1991.

Kestner, Joseph A. *Mythology and Misogyny: The Social Discourse of Nineteenth-Century British Classical-Subject Painting*, Madison: University of Wisconsin Press, 1989.

Kockel, Valentin. "I modelli di Pompei dal Settecento al 'grande Plastico'. La documentazione delle antiche rovine." In *Pompei e l'Europa: atti del convegno: Pompei nell'archeologia e nell'arte dal neoclassico al post-classico*, edited by Massimo Osanna, Rosanna Cioffi, Almerinda Di Benedetto, and Luigi Gallo, 267–75. Milan: Electa, 2015.

Kovacs, Claire. "Pompeii and Its Material Reproductions: The Rise of a Tourist Site in the Nineteenth Century." *Journal of Tourism History* 3.1 (2013): 25–49.

Landow, George. "Victorianized Romans: Images of Rome in Victorian Painting." *Victorian Literature and Culture* 12 (1984): 29–51.

Leardi, G., ed. *I Bazzani a Pompei: I Disegni e acquarelli nell'archivio di stato di Terni*. Terni: Archivio di Stato, 2016.

Lippincott, Louise. *Lawrence Alma-Tadema: Spring*. Malibu: The J. Paul Getty Museum, 1990.

Lowenthal, David. *The Past is a Foreign Country.* Cambridge: Cambridge University Press, 1985.

Macaulay-Lewis, Elizabeth."Imported Exotica: Approaches to the Study of the Ancient Plant Trade," *Bollettino di Archeologia* D9 (2010). www.archaeologia.beniculturali.it/pubblicazioni.html.

Maiuri, Amedeo. *Pompei, con 14 acquerelli originali di Luigi Bazzani e 193 fotografie.* Novara: Istituto geografico de Agostini, 1928.

Miraglia, Marina. "La fotografia, Pompei e l'Antico. Fra documentazione, stile 'documentario' e tensioni estetiche." In *Pompei. La Fotografia,* edited by Marina Miraglia and Massimo Osanna, 31–55. Milan: Electa, 2015.

Monkhouse, Cosmo. "Laurens Alma-Tadema, R.A." *Scribner's Magazine* 18 (Dec. 1895): 663–80, 672.

Monnier, Marc. *The Wonders of Pompei.* New York: C. Scribner & Company, 1871; 1886.

Moormann, Eric M. *Pompeii's Ashes: The Reception of the Cities Buried by Vesuvius in Literature, Music and Drama.* Berlin: De Gruyter, 2015.

Moser, Stephanie. *Painting Antiquity: Ancient Egypt in the Art of Lawrence Alma-Tadema, Edward Poynter and Edwin Long.* Oxford: Oxford University Press, 2019.

Niccolini, Fausto and Felice Niccolini. *Le case ed i monumenti di Pompei disegnati e descritti.* Vol. 4.1, Naples: Fausto Niccolini, 1854–96.

Niccolini, Fausto and Felice Niccolini. "Casa di Marco Lucrezio." In *Le case ed i monumenti di Pompei. Disegnati e descritti,* Vol. 1, Naples: Fausto Niccolini, 1854.

Nichols, Kate. *Greece and Rome at the Crystal Palace: Classical Sculpture and Modern Britain, 1854–1936.* Oxford: Oxford University Press, 2015.

Nora, Pierre. "Between Memory and History: Les Lieux de Mémoire." *Representations* 26 (1989): 7–24.

Nwokobia, Mavis. "Saying It with Flowers: Alma-Tadema and the Flora of Pompeii." *The British Art Journal* 5.3 (2004): 56–61.

Nwokobia, Mavis. "Picturing Pompeii: Lawrence Alma-Tadema and the Legacy of Vesuvius 1863–1912." Dissertation, University of Manchester, 2006.

Pagán, Victoria E. *Horticulture and the Roman Shaping of Nature.* Oxford Handbooks Online, 2016. https://www.oxfordhandbooks.com/view/10.1093/oxfordhb/9780199935390.001.0001/oxfordhb-9780199935390-e-78#oxfordhb-9780199935390-e-78-bibItem-50.

Page, Judith W. "Gardening for Women: Frances Garnet Wolseley and the Rise of the Professional Woman gardener." In *Disciples of Flora: Gardens in History and Culture,* edited by Victoria Emma Pagán, Judith W. Page, and Brigitte Weltman-Aron, 51–67. Newcastle upon Tyne: Cambridge Scholars Publishing, 2015.

Page, Judith W. and Elise L. Smith, eds. *Women, Literature, and the Domesticated Landscape: England's Disciples of Flora, 1780–1870,* Cambridge: Cambridge University Press, 2011.

Piccoli, Chiara. "Visualizing Antiquity before the Digital Age: Early and Late Modern Reconstructions of Greek and Roman Cityscapes." *Analecta Praehistorica Leidensia. Excerpta Archaeologica Leidensia* II, 2017.

Pohlmann, Ulrich. "Alma-Tadema and Photography." In *Sir Lawrence Alma-Tadema,* edited by Edwin Becker, 11–27. New York: Rizzoli, 1997.

Pollock, Griselda. "Modernity and the Spaces of Femininity." In *Vision and Difference: Feminism, Femininity and the Histories of Art,* 70–127. New York: Routledge, 1988.

Prettejohn, Elizabeth. "Recreating Rome in Victorian Painting: From History to Genre." In *Imagining Rome British Artists and Rome in the Nineteenth century,* edited by Michael Liversidge and Catharine Edwards, 54–69. London: Merrell Publishers Ltd, 1996.

Prettejohn, Elizabeth. "Antiquity Fragmented and Reconstructed. Alma-Tadema's Compositions." In *Sir Lawrence Alma-Tadema. Catalogue Van Gogh Museum and Walker Art Gallery, Amsterdam and Liverpool*, edited by Edwin Becker and Elizabeth Prettejohn, 33–44. New York: Rizzoli, 1997.

Prettejohn, Elizabeth. "Lawrence Alma-Tadema and the Modern City of Ancient Rome." *The Art Bulletin* 84.1 (2002): 115–29.

Prettejohn, Elizabeth and Peter Trippi, eds. *Lawrence Alma-Tadema: At Home in Antiquity*, English ed. Munich: Prestel, 2016.

Rambaldi, Simone. "Echi pompeiani ed ercolanesi nella scenografia teatrale dels XIX secolo." *Rivista di Studi Pompeiani* 30 (2019): 61–9.

Rovira-Guardiola, Rosario. "Archaeology in Alma-Tadema's Painting: The Influence of Pompeii." In *The Legacy of Antiquity: New Perspectives in the Reception of the Classical World*, edited by Lenia Kouneni, 161–81. Newcastle upon Tyne: Cambridge Scholars Publishing, 2014.

Ruskin, John. *The Art of England: Lectures Given in Oxford*. New York: John Wiley & Sons, 1883.

Sambuco, Patrizia. *Italian Women Writers, 1800–2000: Boundaries, Borders, and Transgression*, London: Rowman & Littlefield, 2015.

Sampaolo, Valeria. "La realizzazione del plastico di Pompei." *Il Museo* 3 (1993): 79–95.

Savani, Giacomo. "Sensing the Past: Sensory Stimuli in Nineteenth-Century Depictions of Roman Baths." In *The Smells and Senses of Antiquity in the Modern Imagination*, edited by Adeline Grand-Clément and Charlotte Ribeyrol, 119–37. London: Bloomsbury Academic, 2022.

Scagliarini, Daniella, Antonella Coralini, and Riccardo Helg, eds. *Davvero!: la Pompei di fine '800 nella pittura di Luigi Bazzani.* Bologna: Fondazione del Monte, 2013.

Schouw, Joakim F. *Die Erde, die Pflanzen und der Mensch* (1851). Whitefish, MT: Kessinger Publishing, LLC, 2010.

Sogliano, Antonio. *La casa dei Vettii in Pompei. MonAnt* 8 (1898): 233–388.

Stefani, Grete and Michele Borgongino. "I Giardini Pompeiana tra Realtà e Finzione." In *Mito e Natura. Dalla Grecia a Pompei. Catalogo della mostra (Milano 22 luglio 2015–10 gennaio 2016),* edited by Gemma Sena Chiesa and Angela Pontrandolfo, 29–31. Milan: Electa, 2016.

Swanson, Vern. *The Biography and Catalogue Raisonné of the Paintings of Sir Lawrence Alma-Tadema.* London: Garton & Co Editions in association with Scholar Press, 1990.

Urry, John. *The Tourist Gaze: Leisure and Travel in Contemporary Societies,* London: SAGE Publications Ltd, 1990.

Verhoogt, *Art in Reproduction: Nineteenth-Century Prints after Lawrence Alma-Tadema, Josef Israëls and Ary Scheffer,* 427–95. Amsterdam: Amsterdam University Press, 2007.

Verhoogt, Robert. "Reproducing Alma-Tadema." In *Lawrence Alma-Tadema: At Home in Antiquity*, edited by Elizabeth Prettejohn and Peter Trippi, 166–7. Munich: Prestel, 2016.

Zevi, Fausto. *Pompei 1748–1980. I tempi della documentazione*. Rome: Multigrafica, 1981.

Zimmern, Helen. *Sir L. Alma-Tadema, Royal Academician, His Life and Work.* London: Art Journal Office, 1886.

Zimmern, Helen. *Sir Lawrence Alma Tadema.* London: George Bell & Sons, 1902.

3 The Garden's Transformational Artifice in Valois France

Elizabeth Ross

The imagined space of the garden pervades medieval literature, but also related pictorial arts, as one of the most emblematic spaces of courtliness. To take a famous example, the calendar pages at the beginning of the Très Riches Heures, painted by the brothers Herman, Paul, and Jean de Limbourg around 1412–16, juxtapose agricultural labor with the structured and stylized pursuits that mediated courtiers' encounters with nature.[1] They also juxtapose two forms of collaboration, the courtiers' social performance and the laborers' collaboration with the seasons and natural elements. Created for Jean, Duke of Berry, brother of King Charles V of France, each of the pages sets that contrast of peasant and courtier in the frame of the duke's demesne, usually shown through a portrait of one of his residences.[2] For April, the laborers tend fishing nets in the background at the foot of the Château de Dourdan, as a foil to the nobles who pick flowers at what seems to be a betrothal ceremony in the foreground.[3] The courtiers stand outside a walled garden, doubly framed by the garden and the fruit orchards, pasture and timberlands of the surrounding estate. The demesne's even rows of trees, some pollarded, show the same regimentation as the rows of espaliered trees and vines gridded to trellises in the garden proper.

The elite's cultivation of behaviors and virtues—the cultural collaboration called courtliness—conformed to a social code, distinguished members of the landed upper class from others, and required disciplining the body by disciplining desire and gesture. This social code is made visible through dress and other visual markers on and around the body. The creation and maintenance of a garden parallels this process; a garden is a zone, where the natural world is disciplined by artifice and dressed by cultivation just as the natural body is shaped by courtliness. In the adjacent zone of the demesne, the lord cultivates the landscape (through the labor of his peasants) as both the source of his wealth and sign of his dominion.

This bleeds into an adjacent political metaphor where the territory flourishes as a delightful and beautiful place, blessed by God who ordained its rulers, and is richly suggestive of heavenly Paradise.[4] The *Songe du vergier* (Dream of the Orchard), for example, a political tract prepared for Charles V, sets the author's dream of a dialogue between a knight and a cleric in an "orchard" full of roses, lilies, trees, and other delights, made beautiful because the king is enthroned there.[5] The frontispiece miniature from the copy presented to Charles V in 1378 (British Library

DOI: 10.4324/9781003381549-4

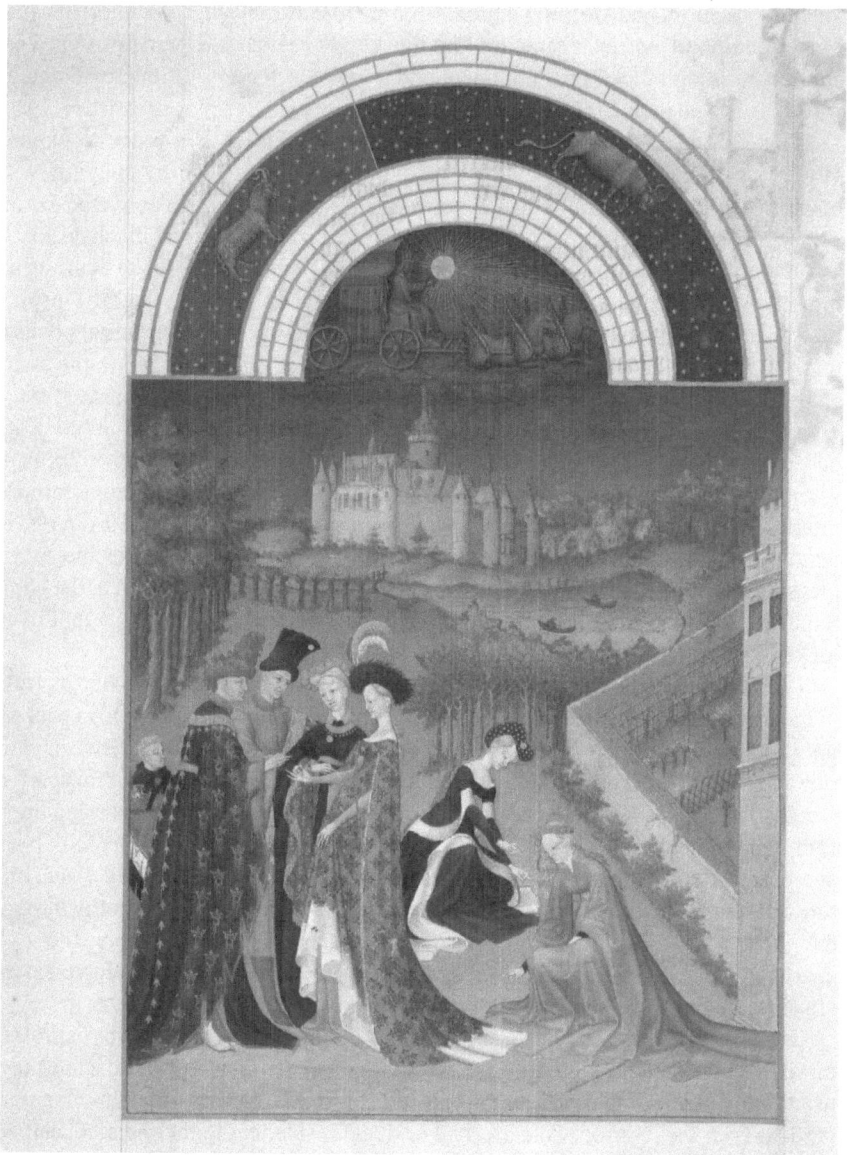

Figure 3.1 Herman, Paul, and Jean de Limbourg, *April* from the *Très Riches Heures du Duc de Berry*, circa 1411/12–16, Chantilly, Musée Condé, MS 65, fol. 4v., Photo: © RMN-Grand Palais/ Art Resource, NY.

MS 19 C IV, fol. 1v.) visualizes this metaphorical apparatus as a garden—an oblong plot of grass, flowers, and a few trees enclosed by a double loop of serried trees. The king sits enthroned at the top of the enclosure at the center of a trinity of trees, between seated queens representing spiritual and temporal power, and above the knight, cleric, and sleeping author.[6] In a treatise establishing the supremacy of French temporal power, characterizing the king's realm as a garden offers both another building block in the argument and a mode of flattery for the royal reader, who sees in this image the delightful results of his rule.

In late medieval literature and pictorial arts, the garden is a space to explore virtue, sex, gender, group identity, technology, and wonder by playing with boundaries—by testing the balance between the innocence of Eden or the *hortus conclusus* and the fall into sin and exile.[7] Or, by using manmade mechanisms to simulate life, to entangle art and nature in uncanny ways that provoke admiration, most famously at the garden of Hesdin in Artois or in Hesdin's literary inspirations.[8] The garden is a ludic space that uses its liminality—its conceptual position on the edge between art and nature and its physical position attached to the architecture of court but outside—to constitute the boundaries of courtliness. It plays both sides of the threshold between natural and artificial, rustic and noble, Eden and exile. In the early Valois courts, this liminality could take another form by visually connecting interior court space with exterior gardens or by transforming interior space into a garden through wall painting or tapestry. At the Louvre, the length of the king and queen's royal apartments afforded a view over the palace pleasure garden to the north, and their apartments abutted them as well at the Hôtel Saint-Pol with their two separate living spaces connected by a gallery that overlooked the garden along its length.[9]

The garden does not just spread—conceptually and physically—beyond its walls in the outside environment into the landscape of the demesne; it could also pervade interior space. This diffusion inside and outside and the connection between exterior and interior offers a model for placing the better-known painting of this period, works like the Très Riches Heures, in the context of a more all-encompassing transformation of court space. Such play with the transformation of interiors took place at a moment when Charles V was beginning to develop the rituals and visual culture of French court ceremony that would distinguish the French monarchy through the *ancien régime*. Accompanying developments in architecture, such as the royal apartments reached by a spiral staircase at the Louvre, a feature also adopted at the Hôtel Saint-Pol and elsewhere, facilitated the new forms.[10]

In the Valois courts in France and the Low Countries, courtliness was expressed through a particular kind of artifice—simulation or "semblance" to use a term they used; and these simulations could take the form of gardens—the semblance of gardens—on the interior walls of royal or ducal residences, either painted, carved, or woven in sets of tapestries.[11] In this way, they suggest the ruler's demesne as a cultivated garden and an earthly paradise. The genre uses a medium of transformational artifice: painted or tapestry interiors that dress a room in the semblance of something else. The genre takes as its subject another kind of transformational artifice—the cultivation of nature. And it does this to create a social space determined by yet another kind of transformational artifice—the artifice that creates courtliness.

As in the Très Riches Heures, the transformational artifice of the garden, demesne, and courtier can use rustics as a foil or a screen for the projection of ideas about elite identity. At the very beginning of the culture of courtliness in the twelfth century, Andreas Capellanus colorfully caricatured courtly love by contrasting it with the raw sex acts of peasants, who make love naturally, like animals without reason. He specifies "like a horse or a mule" to evoke Tobit 6:17 (DV), where the angel Raphael describes those who "shut out God from themselves, and from their mind, to give themselves to their lust, as the horse and mule, which have not understanding …." He cautions against introducing them to the culture of courtly love, lest they neglect the farm work useful to courtiers.[12] That juxtaposition of cultured courtier and natural peasant continues in poetry written in the circle of the court of Charles V and his son Charles VI in the later fourteenth century. In one popular genre, the *pastourelle*, a noble man propositions a shepherdess in a rural encounter, and the verse explores the ribaldry and violence of a situation with a female interlocutor without the protection of class formalized by the artful conventions of courtly love. Both Jean Froissart and Christine de Pizan adapted the *pastourelle* to their political and social purposes.[13] Other poems evoke a related, but different genre, the pastoral, that uses a conversation with shepherds or other rustics to reflect on the qualities that determine urbanity. In verses by Eustache Deschamps, for example, the ambling or traveling poet overhears peasants conversing about the ongoing Hundred Years War, and their words are used as a vehicle for critical political commentary, a version of another poetic format where he laments the corruption of court life.[14]

The first two calendar pages from the Très Riches Heures represent New Year's festivities in January followed by February snow, but they do not share an opening. February's rustic interior, with peasants seated before a fire, follows in series with the turn of the page, after the courtly interior with the Duke seated before his hearth. The juxtaposition proposes the February dwelling as the antithesis or as a travesty of January's courtly hall. Although the modest house is more commodious than the densely packed sheep pen, the shelters rhyme compositionally to suggest a comparison: the artist has arranged our view so that the house runs parallel to the pen and likewise sits open to the elements. A pack of drab birds crowd a smattering of grain on the cold ground after a vibrant party of attendants cluster around the duke's table. This relationship is telegraphed by a two-part dirty joke shared between the pages: a quick groan over the visual pun of the dagger hilts at the waists of the carver and wine steward at the New Year's feast becomes something else as we espy naked genitals stripped of the cover of metaphorical play as the peasants lift their clothes to warm themselves underneath. Michael Camille has made much of the ribaldry in these illuminations to explore the coincidence of sexual and acquisitive desire in the collecting of Jean de Berry, a reading that establishes a sexual charge in the manuscript's intended reception.[15] The relationship between the tandem interiors can be inflected differently— reading the turn of the page as a peeling back of the transformational artifice that creates court space, court bodies, and court objects. It is not a simple flip from concealed to revealed, clothed versus exposed. The body parts of the hovel are revealed to us also at the feast. But there at the New Year's table, courtliness resides in the presentation of one simple thing— the raw material of the body— in the semblance of something else.

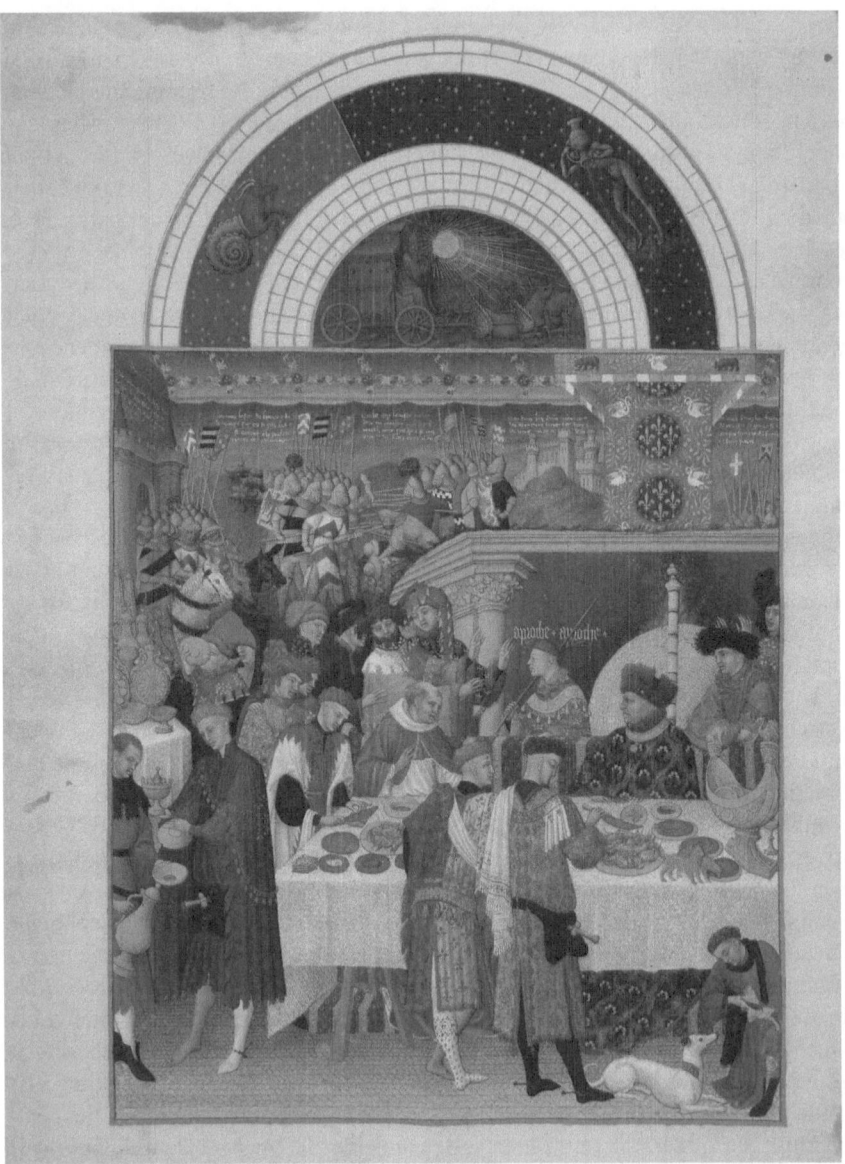

Figure 3.2 Herman, Paul and Jean de Limbourg, *January* from the *Très Riches Heures du Duc de Berry*, circa 1411/12–16, Chantilly, Musée Condé, MS 65, fol. 1v., Photo: © RMN-Grand Palais/ Art Resource, NY.

The foil of February throws into relief how the New Year's scene works to differentiate the nature of court space. The miniature's frame hermetically seals out the winter world of agriculture and animal husbandry, while the tapestry wall coverings frame the feast within a different landscape, now a fair-weather panorama

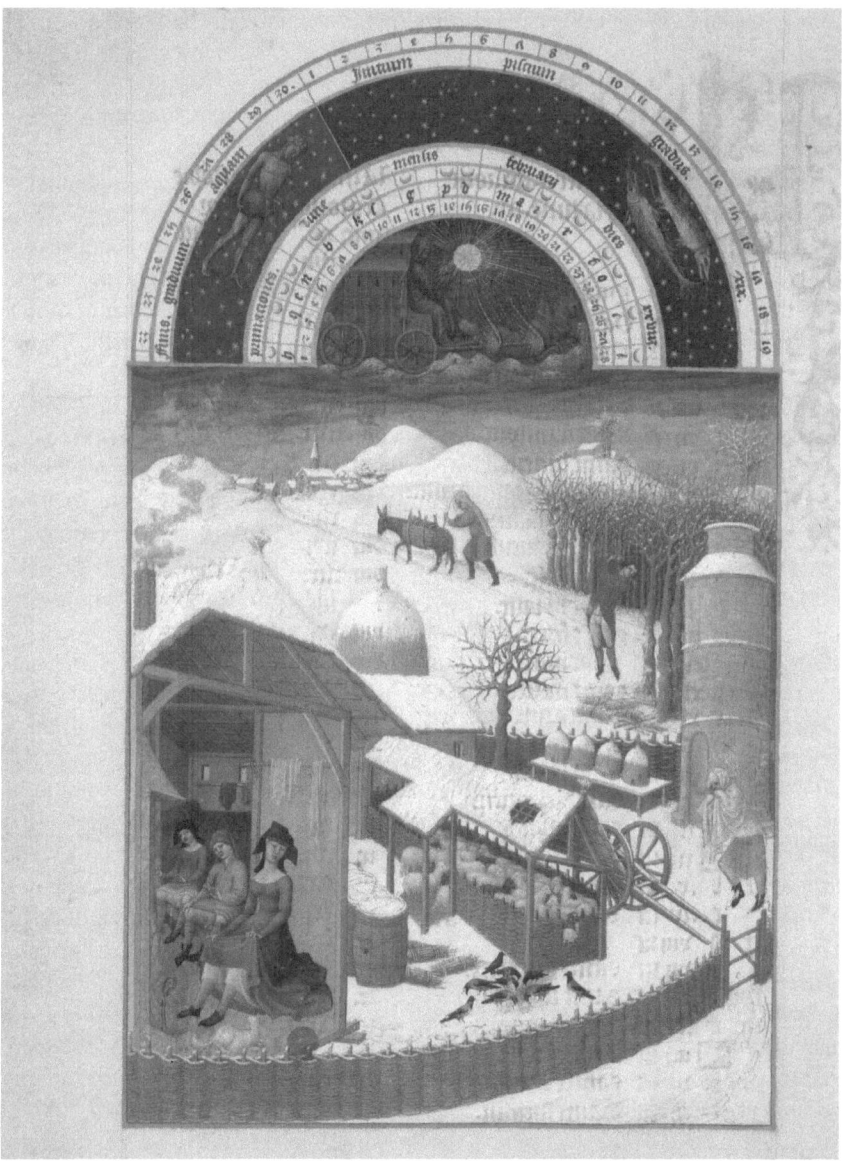

Figure 3.3 Herman, Paul and Jean de Limbourg, *February* from the *Très Riches Heures du Duc de Berry*, circa 1411/12–16, Chantilly, Musée Condé, MS 65, fol. 2v., Photo: © RMN-Grand Palais/ Art Resource, NY.

that may be the Trojan War cast as knightly pursuits proper to their class.[16] The artists underscore both the strength of the transformation of the room and the illusion's material support, by allowing the tapestries' pictorial field to merge entirely with the pictorial space of the page and by clearly delineating the hooks that hold the woven panel. In contrast, the peasant's shelter is materially bare and porous to the elements, suggesting it and its inhabitants' contiguity with basic earth. It is also bare of the layers of signification, largely carried by textiles but also by precious metalwork and social performance, that set apart the space of courtly activity and frame it as a special domain, strongly inflected by courtly ideals.

These calendar scenes intend this contrast to present the peasants pejoratively against a celebration of the collaborative artifice of the court. Equating peasants to animals and raw nature can be a crude joke, but the flipside of that comparison points to a harmony, a genuine collaboration between the peasants and the natural word that is denied the courtly classes. The more courtiers play with artifice or rely on it to distinguish their station, the more isolated they become from artifice's opposite, the natural world left to the peasants.

The Valois courts are full of this kind of transformational artifice. In the New Year's exchange of gifts in 1411, the artists of the *Très Riches Heures* presented their patron the Duke Jean de Berry with a work described as a "counterfeit" made from a piece of wood painted in the "semblance" of a book, but without pages or script.[17] This most art-historically famous of holiday gifts is an old chestnut in the literature for several reasons: it implies an intellectual collaboration between court artist and patron; the joke plays self-consciously off the typical staging of an author's presentation of work to a patron; and it provides an exemplary model for the disengagement of the value of art from the value of an artwork's material substrate.[18] While the artists' success in imitating a book demonstrates their technical mastery, these other aspects of the meaning of the work arise from its insertion in the ceremony of gift exchange, a collaboration that simultaneously helps establish and make visible the social space of the court.[19] After all, the faux book would be just another polychromed sculpture without the successful attempt, drawing on the patron's goodwill and special appreciation for the artists, to substitute wit and surprise for the precious metals that usually supply a gift's value. The Limbourg Brothers' book takes as its theme how transformation-through-artifice circulates to create and distinguish court space, and it demonstrates how this arises as a social collaboration among courtiers in different roles.

There are tapestry inventories that describe sets like the one seen in *January* in the collections of Jean de Berry and his royal brothers and sisters-in-law. Yet, there are few extant works that dramatize for us the dressing of interior space through sculpture, tapestries, or wall painting into a faux environment of courtly cultivation. In Paris, one visible example is the spiral staircase in the tower of the Hôtel d'Artois, the Parisian residence of the Dukes of Burgundy and Flanders. It was built for Duke John the Fearless, nephew of Jean de Berry and Charles V, from 1409–11 at about the same time as the Limbourgs began the *Très Riches Heures*. It was inspired by a staircase built by Charles V in his remodeling of the royal palace of the Louvre. The term *vis* for these spiral stairs is based on an organic metaphor derived from the Latin *vitis* (meaning "vine") to evoke the winding tendrils that propel a climbing plant. The vault of John the Fearless's tower realizes that etymology by reinterpreting the architectural structure as a canopy of leafy oak and hawthorn tree limbs wound with hop vines. These are heraldic plants representing

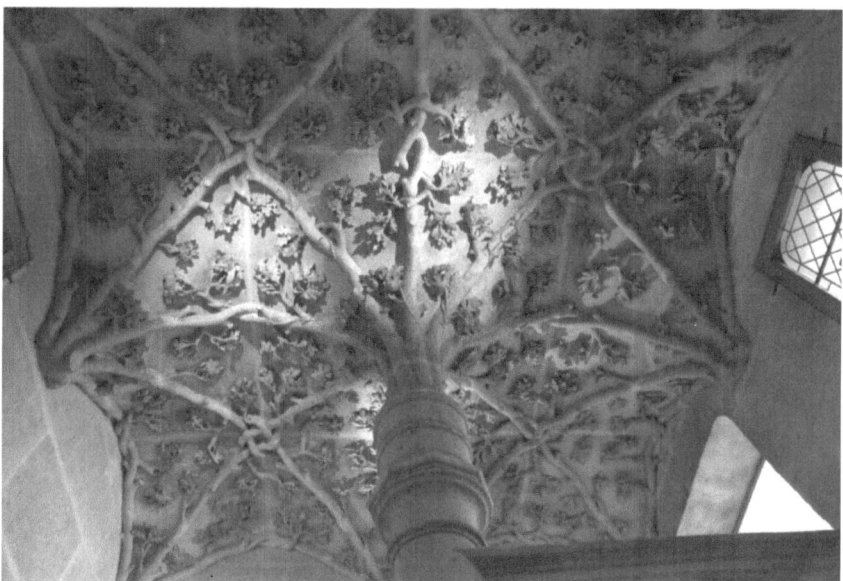

Figure 3.4 Claus de Werve, Vault of the spiral staircase in the Tower of John the Fearless, 1409–11, Paris, Photo: Elizabeth Ross.

the current duke and his parents. John the Fearless had used the hop vine and its seed cone as a device from his youth before he ascended to the dukedom, likely in recognition of the economic role of beer, increasingly brewed from hops, in the prosperity of Flanders. The oak symbol of his father, Philip the Bold, and the hawthorn of his mother, Margaret of Flanders, support his hop vine as it grows and ascends through the structure of the trees, winding up like the staircase. The heraldic symbolism adds a political and family tree element that complements the larger political program of the duke's rebuilding of the hôtel and this staircase in triumph after a conflict with the crown.[20] The stairs also demonstrate both the full-bodied meaning inherent in the graphic shorthand of a heraldic device and the period interest in a kind of transformational artifice that gives that meaning greater representational life.

The spiral staircase in the tower of the Hôtel d'Artois survives, but for other royal spaces we are dependent on archival records and the description of the mid-seventeenth-century historian Henri Sauval who studied account documents since lost to fire. Sauval describes an even more elaborate, painted version of this kind of artifice created during the reign of King Charles V:

But in past centuries, there was none more magnificent than that which Charles V completed in the apartment of the queen at the Hôtel Saint-Pol. From the paneling to the vault was represented on a green ground, and beneath a long terrace that extended all around, a grand forest full of trees and shrubs, apple trees, pear, cherry, plum, and others like those, full of fruit, entangled with lilies, yellow flag irises, roses, and all sorts of other flowers, and with children spread among several places in the wood, gathering flowers there and eating fruits, the others pushing their branches to reach up into the vault, painted white and blue to represent the sky and the day—cheerful, made of orpiment [yellow pigment] and fine indigo.

He then describes another vignette of transformational artifice: the vault of a covered passage to the queen's oratory at the Church of Saint Paul, painted with angels singing and playing instruments as they descend from a blue heaven.[21]

Sauval's description of the queen's interior inevitably calls to mind an even earlier example, the Chambre du Cerf (Stag Room), the study in the apartment of Clement V in the Palais des Papes in Avignon. Though in poor condition further marred by an infelicitous 1907 restoration, the decoration of Clement V's study has survived well enough to help envision what these painted environments might have looked like. In 1342–43, a team of French and Italian artists painted the walls in a tower room with a wrap-around landscape populated by young men, boys, and one older man engaged in various forms of hunting, fishing, and bathing. The papal painting seems to have influenced the Valois courts directly through the close ties

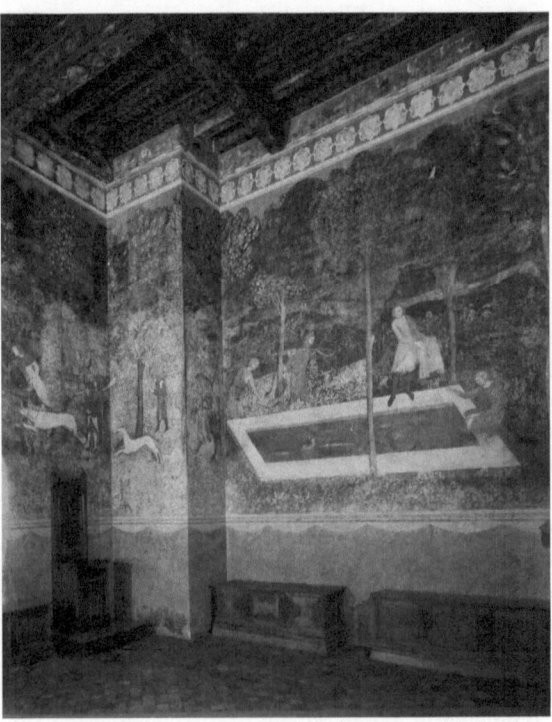

Figure 3.5 Detail from Chambre du Cerf (Stag Room), 1342–43, Avignon, Palais des Papes, Photo: Erich Lessing / Art Resource.

of Clement V and Charles V's father, King Jean II the Good. Jean II made three visits to Avignon, including a stay from December 1350 to January 1351, a few months after his accession, with his court artist Girard d'Orléans in tow. Shortly before his coronation, Jean II began a major campaign of decoration at the castle of Le Vaudreuil in Normandy (now destroyed), his seat as duke of Normandy before taking the throne. Girard d'Orléans began the work before turning it over to Jean Coste after the new king and his artist relocated to Paris, although Girard continued as supervisor.[22] As the influence of the Avignon School's distinctive style has been identified elsewhere in works produced for Jean II and his son, scholars have speculated that the "hunt" painted on the walls of a gallery outside the Great Hall of Le Vaudreuil found its inspiration in the Chambre du Cerf.[23] This line of influence connects to the queen's apartment at Saint-Pol and also assists in explaining Charles V's decoration of the Great Hall of the Louvre in a similar manner in 1366. Sauval relates that the Louvre painting was renewed in 1514 to emphasize "birds and animals that played in large country landscapes accompanied by stags."[24]

In the Chambre du Cerf a thicket of trees forms the background, blocking any vista and limiting the recession into space, with the open area in front carpeted by low vegetation punctuated with a few trees. The human activity largely takes place in the middle-ground: hunting with dogs for deer; flushing rabbits with a ferret; fishing with nets and hooks in a man-made pond; stalking birds across multiple vignettes with a falcon, owl, bird call, nets, traps, and the climbing of trees; and children bathing in a river. Dieter Blume has recently drawn attention to the absence of the usual trappings of aristocratic hunting, for example courtiers on horseback, and the seeming ineffectiveness of many of the hunting strategies.[25] He proposes to read the mural in a Petrarchan vein, using a depiction of nature as a caution against getting mired in the vain pursuits of this world, rather than uplifting your minds (*animi*): "endeavor not to catch a bird, but to become one."[26] Do not get mired in the artifice of cultivated spaces in a way that isolates you from the inspiration of Creation. Earlier readings have sought an allegorical explanation for the hunting imagery also appropriate to a pope, and they have noted, as Blume does, that the profusion of identifiable species would have provided the pleasure of recognition to the knowledgeable patron.[27]

The panoramic organization of space with a forested background and pockets of activity on a verdant fore- and middle-ground stood in dialogue with an emerging genre of tapestries where courtiers exercise the exclusively noble prerogative to pursue game on their land, while also pursuing each other through the rituals of courtly love.[28] The earliest extant panel fragments with secular scenes have been dated to 1400–10. All the activities take place in small clearings among trees or on a grassy area with flowers in front of trees: one shows an apparent betrothal in front of a castle as if a condensed version of the Limbourg's *April*, two depict scenes of falconry and conventions of courtly love, another offers a courtly love scene, and a fifth displays a boar hunt and courtly ladies picking flowers and making music.[29] The *Bear and Boar Hunt* in the Victoria & Albert Museum, dated to 1425–30 on the basis of fashion, offers an example that has survived largely intact with dimensions of about 3.8m in height by over 10m in length. *Boar and Bear Hunt* belongs to a group, the Devonshire Hunting Tapestries, that entered the museum together, though differences in style, materials, and workmanship indicate they came from different sets produced through 1450. Together they depict falconry, a deer hunt, and a hunt for a swan and otters.[30]

Readings of the *Chambre du Cerf* underscore, however, that transformed interiors from this era did not simply anticipate later genres, such as hunt tapestries, especially when the space has been designed for a patron like a cleric or queen. The "hunt" recorded at Le Vaudreuil or the country landscape with stags at the Louvre may describe content more like the Chambre du Cerf than the Devonshire panels. For example, the tapestries that the brother of Charles V and Jean de Berry gifted the Emperor Sigismund of Hungary in 1416 depicted "lords and ladies hunting birds" and "small children taking birds." These have been cited to demonstrate that there were tapestries like *Bear and Boar Hunt* that predated the extant examples, but the description even more aptly evokes the boys chasing birds in trees of the Chambre du Cerf or the focus on children in the queen's apartment at Saint-Pol.[31] Charles VI's wife, Isabeau of Bavaria, bought a tapestry with men and women fishing and pursuing "several pleasurable activities," while an inventory at the death of Margaret of Flanders, recorded at least two tapestries with children playing, one that also included adults pursuing amusements in a grassy area (*herbette*).[32]

With the destruction of the royal interiors described by Sauval, the search turns to echoes of such schemes in other media. A bas-de-page from the Turin-Milan Hours (Turin, Museo Civico d'Arte Antica, inv. no. 47, fol. 113r.), one of the most famous and scrutinized manuscripts of the period, depicts a celebratory feast with the king's table and canopy of honor set before an interior wall decorated with a gilded garden trellis entwined with rose canes or similar leafed and flowering vines. The three images on the page—main miniature, illuminated initial, and bas-de-page—work together to gloss the king's hall as a garden that prefigures heavenly paradise, and iconographically related tapestries carried the same ideas off the page into the architectural space of the court.

The Turin-Milan Hours began in 1389 as a commission by Jean de Berry in Paris. The first sixteen gatherings were split off as a complete and independent Book of Hours that passed to the duke's treasurer. The history of the rest has been the subject of tremendous debate, catalyzed by a theory that several pages of the manuscript were created by a young Jan van Eyck who brought the manuscript to Bruges after his first patron's death. Scholars have vigorously debated the date and number of campaigns of illumination and the attribution of pages to various hands.[33] The bas-de-page discussed here does not come from one of the pages with contested authorship or dating. It was produced in Bruges after van Eyck's death in the later 1440s or 1450s for someone in the circle of Philip the Good, the Duke of Burgundy and Flanders, perhaps as a gift for him.[34] The passage of the Turin-Milan Hours from Paris to Bruges exemplifies a passing of the artistic baton to the court of Philip the Good, the son of John the Fearless, who elaborated the type of court ceremonial and spectacle developed by his great uncle, Charles V.

The name of the manuscript belies the fact that it also contains a missal. The page discussed here begins the text of the mass for the Feast of All Saints, which is visually introduced by this large miniature of the *Joys of the Blessed* in the court of heaven. The text speaks of celebrating the feast in honor of all the saints (*diem festum celebrantes sub honore Sanctorum onmium*), in which the angels rejoice, while the archangels praise the Son of God (*de quorum solemnitate gaudent angeli et collaudant filium dei*), and the scene in the bas-de-page is included for its typological association with the celestial festivities. Sitting beneath the simulated gold trellis is the Persian King Ahasuerus (also known as Xerxes I) from the Book of Esther. Ahasuerus is best known for his role in the main narrative of the Book of Esther: his Jewish wife and queen, Esther, successfully asks him to intervene to prevent a plot, cooked up by a royal advisor, to exterminate the Jews in Persia. The biblical narrative opens with the scene depicted in the bas-de-page, where Ahasuerus throws an epic feast that lasted 180 days to "shew the riches of the glory of his kingdom, and the greatness."[35] He capped that off with a seven-day feast set up "in the court of the garden, and of the wood, which was planted by the care and the hand of the king," where he erected a lavish pavilion.[36] In the *Speculum humanae salvationis* (Mirror of Human Salvation), the delights and the length of the feast of Ahasuerus, as well as his inclusive invitation, prefigure the eternal delights of the heavenly celebration to which Christ invites everyone who desires salvation. The *Speculum* is an illustrated fourteenth-century Latin text that was translated into the vernacular and circulated widely in the fifteenth century, including at the court of Philip the Good.[37] The Turin-Milan bas-de-pages designed in Bruges often adopt its typologies.[38] Each opening of the *Speculum* illustrates an essential moment from Adam and Eve's Fall, the life of the Virgin, the life and Passion of Christ, or the Last Judgment, and each of these key scenes are glossed with three supporting images of biblical events, objects, or parables that prefigure it.

The garden imagery goes beyond either the biblical or *Speculum* source material, however, to evoke the associations seen previously in the *Dream of the Orchard* between a ruler and the terrestrial paradise of the demesne under his cultivation. First, the trellis works together with the setting in the initial G, a field dotted with wildflowers next to a woodland. An angel swoops down with Revelation 7:3 (DV), the lesson read near the beginning of the All Saints' mass; he calls out, "Hurt not the earth nor the sea nor the trees, till we sign the servants of our God in their foreheads." In other words, do not start the apocalypse until we anoint the elect who will survive it. This takes place in a kind of *locus amoenus* (pleasant place) whose natural beauty is angelically preserved, a place on earth that leads to paradise, and this setting acts as a gloss on the unusual emphasis in the bas-de-page on the king's hall as a foretaste on the route to paradise. Second, while the artist may have been inspired by the biblical feast's location "in the court of the garden, and of the wood," he did not take that as literally as he could have done. Rather than dining in an outdoor pavilion, Ahasuerus presides over a hall, similar to the setting for Jean de Berry's *January* feast. It is the very heart of his court itself that has been transformed by the metaphorical artifice of the trellis.

Ahasuerus and his queen Esther were deployed in other media at the Burgundian court, and Birgit Franke, who has written the most sustained analysis, argues that they were used to evoke the virtues of rulers, namely the magnificence and accompanying magnanimity of the duke, the ideal attributes of a princely married couple, the duke's justice, and the duchess's protection of her people. At the 1468 wedding of Philip the Good's son, tapestries celebrated the new duke Charles the Bold and his bride as Ahasuerus and Esther. Evidence from account documents suggest that Philip the Good had a set of four Ahasuerus and Esther tapestries restored in 1451 before ordering a set of six as a gift for the cardinal of Arras in 1461–62, and other tapestries with this theme are documented from the late fourteenth century.[39]

Remnants of two sets survive to remember the tapestries bought in 1461–62 or displayed in 1468. The panel in the Minneapolis Institute of Art is only a fragment, but it shows a queen's feast, where Esther asks Ahasuerus to intervene to save her people.[40] Behind the queen seems to hang a tapestry-within-the-tapestry

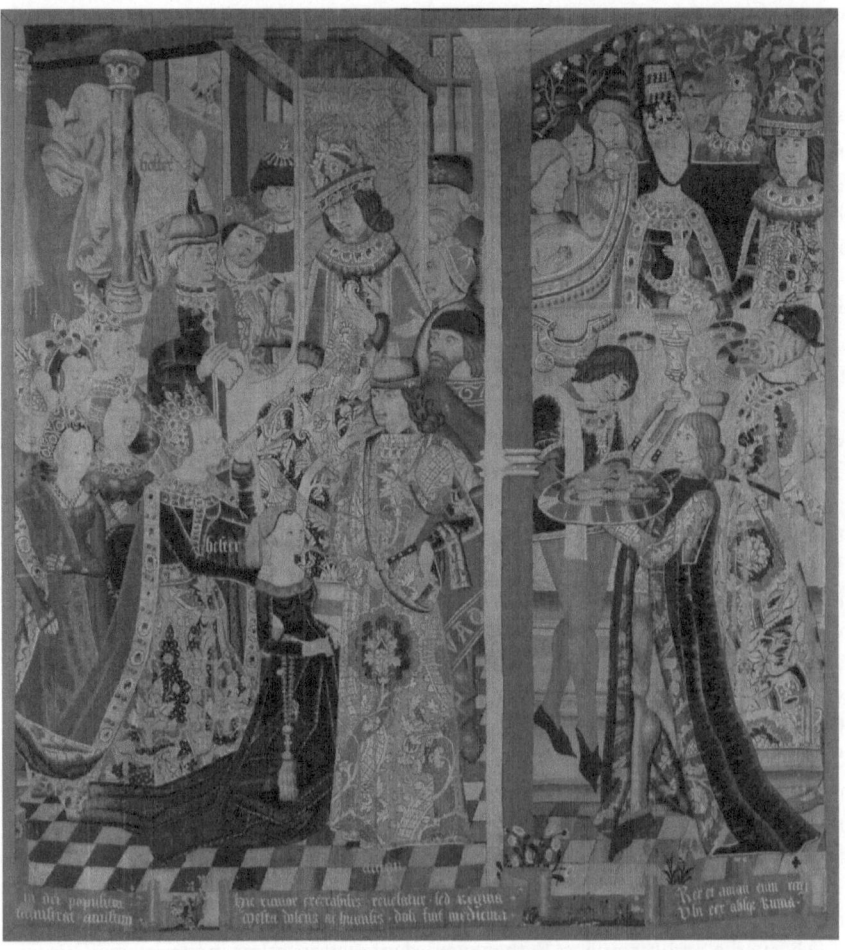

Figure 3.6 Fragment of tapestry with Esther and Ahasuerus, circa 1460–85, Photo: Minneapolis Institute of Art.

showing Esther seated in front of what was called a *verdure* (roughly, greeneries) in the period and now a *millefleur* (thousand flowers) after the dense allover pattern of flowering plants on a green grassy ground. With the awkward borders of a fragment, it is difficult to say for certain whether the greenery behind Esther was intended as a representation of a tapestry or of living greenery, but that ambiguity encapsulates the aim of *verdure* to transform space with greenery. If not inspired by an actual painted ensemble, such as the Great Hall of the Louvre, the painter of the Ahasuerus bas-de-page seems to be artfully reimagining the garden imagery and metaphors of *verdures*.

The finest extant example of *verdure* comes from a set of eight pieces commissioned by duke Philip the Good in 1466. The surviving fragment of one panel depicts the duke's arms, emblem, and monogram against floral ground with at least thirty identifiable species densely arranged to convey a lush ideal. That natural bounty complements the lavish material wealth of the finely knotted silk weaving, where the duke's symbols are composed with thread made from precious metals. Together, exceptional vegetation and materials present the duke's dominion as an Arcadia under his cultivation, a greener and even more luxuriant garden than Charles V's orchard.[41] The dimensions of the original set, known from account documents, suggest it was made for the Coudenberg Palace in Brussels, though

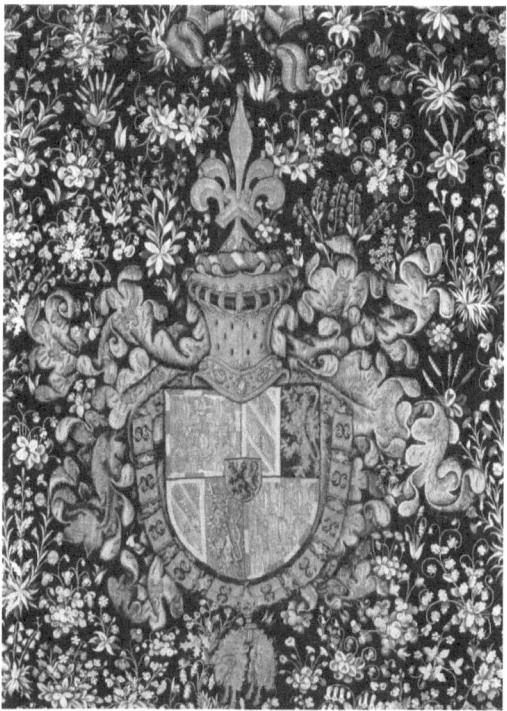

Figure 3.7 Detail from Millefleur tapestry with the arms of Philip the Good, circa 1466, Bern, Historisches Museum, Photo: © DeA Picture Library/Art Resource.

such sets would have traveled with the duke as demonstrated by its subsequent history, when it was captured from Charles's belongings by Swiss mercenaries at the Battle of Grandson in 1476, part of a military campaign that ended in Charles's defeat and death.[42]

Tantalizingly, excavations at Coudenberg have uncovered stone architectural fragments with Philip the Good's emblems—and one with the emblem of a garden enclosed by a wattle fence used by his wife Isabella of Portugal.[43] This suggests husband and wife dressed their court space with heraldry-inflected garden imagery in tandem. With the fenced garden Isabella embraces the chaste wifely virtues signaled by the *hortus conclusus* (enclosed garden), a motif used pervasively as a metaphor for the Virgin—but a metaphor also subject to the ribald play of courtly love literature, where lovers strive to breach the defensive boundaries of the garden. A miniature in a Swiss chronicle shows these tapestries hung as a backdrop to the 1473 meeting in Trier between Charles the Bold and the Holy Roman Emperor Frederick III.[44] This choice likely speaks less to the historical record and more to the artist's local access to the tapestry and his desire to boost their conquest by refreshing its association with Charles the Bold at the pinnacle of his power. Nonetheless, that the illuminator would choose this tapestry to dress such an occasion indicates its potent political messaging.

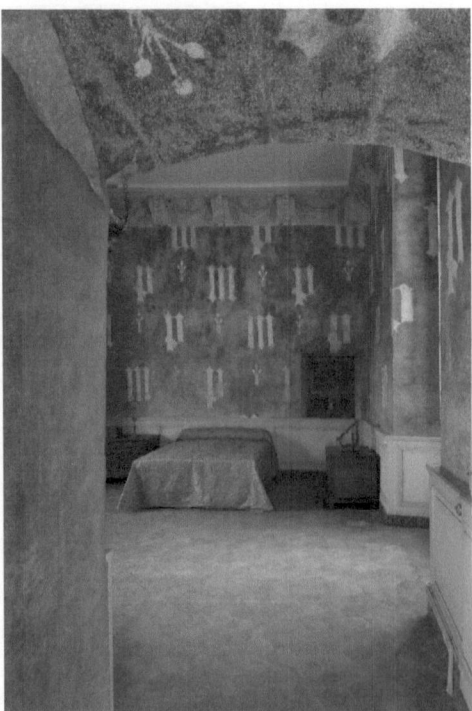

Figure 3.8 Arnoul Picornet, Wall painting in room of Margaret of Bavaria, 1389–90, Château de Germolles, Photo: Manuel Cohen/Art Resource, NY

One example remains, partially intact, to demonstrate the early development of the Valois practice of using transformation-through-artifice to create and distinguish court space—even if it does this without the illusionism of the Chambre du Cerf or Hôtel Saint-Pol (presumed), the naturalist interest of the Chambre du Cerf or Philip the Good's *verdure*, or the wit and ribaldry of the Limbourg's simulated book or phallic daggers. The Château de Germolles in Burgundy, about forty-five miles from Dijon, is the best-preserved fourteenth-century residence from Charles V's circle.[45] Enough survives of the château materially and in archival records for us to be able to reconstruct how the full program at Germolles—estate, gardens, and interiors—worked together to transform the raw nature of the land with the artifice of an ideal demesne. The painted walls at Germolles enact the same conceit as *verdure*, but, even more than the narratively vacant *verdure*, the Germolles interiors rely on emblematic heraldry applied in an allover pattern on the walls and floor in a way that borrows from prestige silk brocade textiles even more than tapestry.

In 1381 Duke Philip the Bold gifted the estate to his wife, the duchess Margaret of Flanders, who undertook significant building and refurbishment over the next decade with the layout of the apartments and the spiral staircase adapting royal architectural models. Margaret developed the property as an agricultural and garden showcase with a vineyard, roses, lavender, lilies, red cabbage, sorrel, sage, pears, cherries, raspberries, gooseberries, several types of cows, including three imported from Flanders, and flock of at least 68 sheep.[46] It was Margaret who brought the wealth of Flanders and nearby northern territories into the marriage and the ducal line. And much of this wealth was derived from taxes on the import of the English wool used to make Flemish cloth and the profit on the trade in finished textiles. Germolles was a model demesne that featured living signs of one of the sources of Margaret's prestige, as well as the kinds of fruits and perfumed plants that belong to the realm of agricultural luxury. The model demesne outside was reiterated and visually crystallized inside through the program of interior decoration in the main apartment. Margaret had the walls and the wood paneling of the ceiling stenciled with her and her husband's initials, motto (*Y ME TARDE*, I am waiting) and symbols—112 heraldic shields, 200 mottos and emblems, 480 sheep, 500 marguerites (a play on her name), 850 thistles, 1013 P and M initials, and 1363 white and red roses.[47] Of these rooms, the room that has been best preserved clearly develops the transformational logic of *verdure* with flowers stenciled in a regular pattern on a green background. The account documenting payment for these motifs assigns the room to Margaret's daughter-in-law, Margaret of Bavaria, wife of John the Fearless.[48]

The interior decoration of the rest of the living apartment has been lost, but the ducal accounts record payment for sheep in addition to flowers, and this motif dominated one space known as "the room of the sheep."[49] The motifs of the walls and ceilings were carried through in the tiling of the floor of the hall in a building across the courtyard from the main apartment. The sheep tiles suggest the form of the sheep emblems that were stenciled on the walls, with sheep taking shelter under an elm.[50] Like all the figurative tiles, the yellow design was set against a red background. Contemporary accounts record purchases destined for this room: red satin textiles richly embroidered with sheep motifs made with gold-wrapped silk thread continue the color scheme of the tiles, and there were additional cloth fittings with the same motif, as well as saddles ornamented with sheep.[51] An inventory of the duchess's belongings records tapestries with a gold tree, flock of sheep, and a shepherd and shepherdess; a shepherd and shepherdess wearing straw hats with sheep, little sheep pens, and the coat of arms of the duke and duchess; and a similar tapestry with gold trees (like the tiles) instead of sheep pens.[52] Such design elements would have transformed that space into a sheep pen or sheep pasture, with the ducal couple, or the duchess and her daughter-in-law who were more often in residence, presiding over the space in person. This imagery launches a long tradition of women in the French royal family surrounding themselves with the idealized trappings of rustic life for pleasure as well as to project ideals of contemporary feminine virtue. A later version of this phenomenon runs from Catherine de' Medici and Madame de Pompadour through to the bitter end of the *ancien régime*, culminating in the Hameau de la Reine at Versailles and the Queen's Dairy at the Château de Rambouillet, both built for Marie Antoinette.[53] Of course, these are forms of transformational artifice, not to be confused with genuine collaboration with nature.

The lynchpin of this pastoral program was a lost sculpture by Claus Sluter at the entrance to the château that depicted the duke and duchess, probably life or near life size, as shepherds with their sheep under an elm. The ducal accounts record its creation, the construction of a structure to shelter it, the refreshing of its polychromy, and the maintenance of the shelter.[54] Sluter is a canonical figure in the art history of this period, appointed in 1389 to head the workshop carrying out the duke's many projects. His three major extant works are found at the duke's Carthusian foundation at Champmol about 2.5km from the ducal seat in Dijon: the portal of the monastery chapel, the Well of Moses at the center of the cloister, and the duke's tomb inside the chapel.[55] At the death of his predecessor, Jean de Marville, he took over the work on the tomb and portal that was in progress, and at the end of his own life, he passed completion of the tomb to his nephew and successor Claus de Werve, the artist believed to have carried out John the Fearless's spiral staircase. The installation of the group at Germolles around 1393 coincides with his work on the portal, his first project at Champmol, which also featured sculpted portrait likenesses of the duke and duchess. Sluter's extant works suggest that his shepherd and shepherdess would have been carried out in a very different style from the emblems in the interior decoration at Germolles, but that stylistic difference belies the sculpture's affinity with the rest of the program. In working together with the emblematic program and the landscape of the estate, the sculpture helps establish that realism is not required to express an enveloping naturalism.

While the Well originally featured a Crucifixion with Mary Magdalene kneeling to embrace the foot of the cross, all that remains are the six prophets mounted around what was the base of the Crucifixion.[56] Some of the rich polychromy applied by Jean Malouel has survived to be restored. As ever, realism is a slippery term that here designates the figures' plausibility as portrait likenesses and

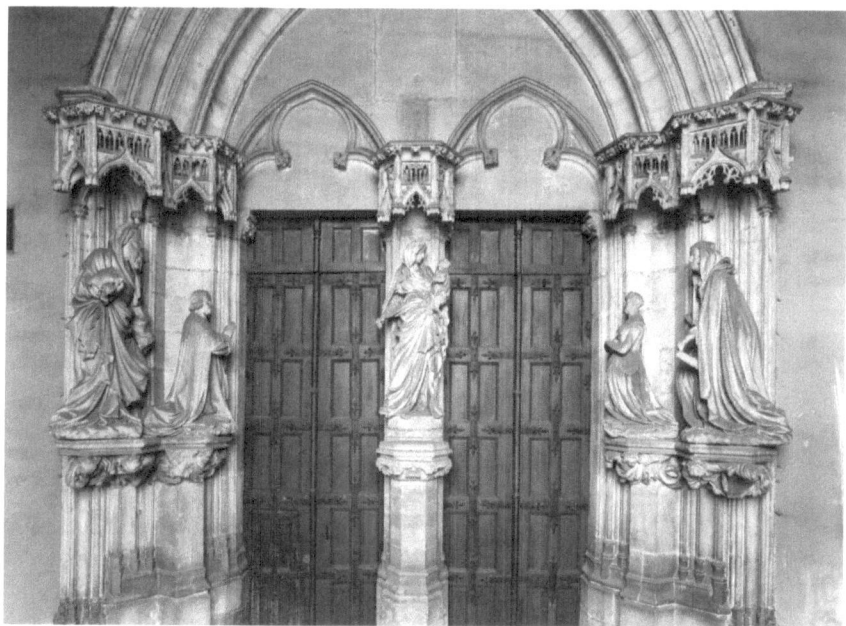

Figure 3.9 Jean de Marville, Claus Sluter, and studio, Portal of the Church of Champmol, 1384–1401, Dijon, Chartreuse de Champmol, Photo: Bridgeman-Giraudon/Art Resource, NY.

character sketches, a kind of convincing historical fiction, as if Sluter had caught six adroit character actors at a moment in their performance that crystallized the different personalities of the prophets. That Jeremiah is recognized as a crypto-portrait of the duke adds to this effect by slipping a recognizable likeness in with fictitious, but equally convincing portraits.[57] In height and bulk, they register with the viewer as human forms, and the figures seem to acknowledge or address each other, leading the viewer around the pedestal, except for the self-contained Zachariah who models the distant gaze of a prophet preoccupied by noncorporeal vision.

This is the hallmark of Sluter's particular realism, his transformational artifice, seen also in the tomb and portal: he creates convincingly human characters who express an array of attitudes and whose illusion is often activated through a parity of scale between the figures and the viewer. On the tomb sculpture, all four sides of the base present an arcuated architectural structure that houses a cortege of mourner figures, mostly Carthusians, about 40cm tall. Each is an individual study in the body language of grief. While here there is no parity of scale between the viewer and figures, the effect of realism is activated kinetically by the viewer's cir-cumambulation of the work, as at the *Well of Moses*. The architectural framework obscures and reveals different aspects of the figures to approximate the shifting sequence of a moving cortege. The portal for the church at Champmol was the first of Sluter's works to use the frame—or rather, the figures' refusal to be contained by the frame—to animate an effect of realism. Full-body portrait likenesses of the duke and the duchess, each on opposite sides of the portal, extend their kneeling pose beyond the traditional socles and baldachins above and below, introduced by their equally expansive standing patron saints who sweep forward to present them to the Virgin and Child in the trumeau.

All three of these works feature portraits (tomb and portal) or crypto-portraits of the duke which are admired for the realism of the likeness. The lost Germolles sculpture teases a fourth work in the same vein. While the record of Sluter's lost Germolles sculpture is known to specialists, its loss means that it has not yet found a place in the narrative that recognizes how it functions like Sluter's other works to transform space. Attention to the way Sluter's figures engage with the space around them—the space of the viewer and the field of movement in front of and around the work—recasts his work from the model of a self-contained sculpture on a pedestal or socle to something more like an installation in our contemporary sense. Focusing on the somatic, kinetic, and spatial strategies in Sluter's work strengthens the presumed connection between the crypto-portraits of the duke and duchess as shepherds and the rest of the estate. The sculpture was one of many complementary aspects of an installation that implicates the entire estate, inside and out.

The sculpture completes the program by populating the transformed ideal demesne with caretakers whose protection and correction ensures the prosperity of their flock. Without the gardens of the estate, the transplanted sheep, the proto-*verdure*, and the rest of the emblems, Sluter's sculpture remains a self-contained metaphor of rulership as pastoral care. Placed in an environment that has been shaped by artifice inside and out, the sculpture completes the transformation of the princely château into a productive *locus amoenus* overseen by skilled guardians following a rustic ideal. The Très Riches Heures presented courtliness as the transformation of the raw natural person into the semblance of something else, the courtier dressed through artifice. So, too, at Germolles the raw rustic has been transformed—into an ideal and artificial version of itself. The joke here is not genitals remade as fine daggers or wood into a book, but shepherds transformed into better versions of themselves in an equally elevated environment.

As a coda, here is another example of the transformation of space with the *Well of Moses*, which stood in the center of a quadrangle bordered by individual dwellings for each monk. The surviving well structure was originally topped by a Crucifixion, which Susie Nash has skillfully reconstructed to explain its devotional purpose.[58] For us, however, the well brings to the center of the Carthusian green space the metaphor of God as the fountain of life, as in, for example Psalm 36:10, a metaphor entangled with interpretations of the blood and water that flowed from Jesus' side when pierced on the cross.[59] This fountain of sacred metaphor was understood as a garden fountain, as depicted, for example, in the Garden of Eden in another illumination for the Très Riches Heures (fol. 25v.) Sluter's *Well* works here to conceptually transform the Carthusians' common space into a garden of God's grace. It is both the promised paradise of salvation, as well as Eden, lost through sin. The monks pointedly did not inhabit either garden. Their discrete shelters edge the perimeter, each with its own small garden plot where they worked the earth with labor and toil to produce food, each like an exiled Adam as described in Genesis 3:17 (DV). In this case, rather than distancing the monks from nature, the transformational artifice of the *Well*, by placing them in a zone of exile from God's ideal garden, instead connects them to the terrestrial natural world.

Notes

1 Alexander, "*Labeur* and *Paresse*."
2 On the artists of the manuscript and their context at the court of the duke, see most recently, Dückers and Roelofs, *The Limbourg Brothers: Nijmegen Masters at the French Court, 1400–1416*. The older standard works are still useful: Meiss, *French Painting in the Time of Jean de Berry: The Late Fourteenth Century and the Patronage of the Duke*, 2nd ed. and *French Painting in the Time of Jean de Berry: The Limbourgs and Their Contemporaries*.
3 For the fashion, van Buren, *Illuminating Fashion*, 122–27.
4 Beaune, *The Birth of an Ideology*, 292–97.
5 The dream takes place "en un vergier qui estoit tres delectable et tres bel, plain de roses et de fleurs de lys et de plusieurs aultres delys, car la vous vis en vostre majesté assiz." Schnerb-Lièvre, *Le Songe du vergier*, vol. 1, 4. Commissioned by Charles V, the Latin version, *Somnium viridarii*, dates to 1376, the French to 1378. The translator of Beaune, *Birth of an Ideology*, renders the title as *Dream in the Pleasure Garden* (79–319 *passim*).
6 For the frontispiece, *Paris 1400: Les arts sous Charles VI*, 51.
7 The example *par excellence* for this play is the *Roman de la rose*. See for example, Kay, *The Romance of the Rose*. For these themes in early Netherlandish painting, see Pearson, *Gardens of Love and the Limits of Morality in Early Netherlandish Painting*. Although the painting, sculpture, and prints discussed by Pearson were produced in the same time and region as the fifteenth-century Valois courts, they were, by and large, not made for courts.
8 Truitt, *Medieval Robots: Mechanism, Magic, Nature, and Art*, 122–40; *Art from the Court of Burgundy*, 160–61.
9 Whiteley, "Le Louvre de Charles V"; "Le Château du Louvre" and "Hôtel Saint-Pol," in *Paris et Charles V*, ed. by Pleybert, 117, 123–24.
10 Whiteley, "Louvre de Charles V," 60–71; "Le Château du Louvre" and "Hôtel Saint-Pol," 113–24; "'La Grande Vis': Its development in France from the mid-fourteenth to the mid-fifteenth centuries," 15–20, 219–22.
11 For an example of the use of "semblance," see below.
12 Capellanus, *Andreae Capellani regii Francorum De amore libri tres*, 235.
13 Smith, *The Medieval French Pastourelle Tradition*.
14 Kendrick, "L'invention de l'opinion paysanne dans la poésie d'Eustache Deschamps," esp. 167–72.
15 Camille, "'For Our Devotion and Pleasure': The Sexual Objects of Jean, Duc de Berry."
16 The Trojan War identification comes from Meiss's endorsement of a reading of an inscription, although the duke does not seem to have owned tapestries with this theme. Meiss, *Limbourgs*, 63–64. Trojan tapestries are documented in the collections of four other Valois. McKendrick, "The Great History of Troy," 44–48, 67.
17 "Item, un livre contrefait d'une pièce de bois paincte en semblance d'un livre, où il n'a nuls fueillets ne riens escript …." Guiffrey, *Inventaires de Jean Duc de Berry, 1401–16*, I:265.

18 Cited, for example, in Meiss, *Limbourgs*, 48, 50, 294; Dückers and Roelofs, *Limbourg Brothers*, 21. Camille, "Devotion and Pleasure," 181 uses it as the launching point for a reading of other puns and puzzles in *January*.

19 For this phenomenon, Buettner, "Past Presents: New Year's Gifts at the Valois Courts, ca. 1400."

20 For this work, *Art from the Court of Burgundy*, 158–59; Plagnieux, "La tour Jean Sans Peur: Une épave de la résidence parisienne des ducs de Bourgogne." For the attribution of the sculpture to Claus de Werve, Viré and Lavoye, "Matériaux et phases de construction."

21 "Mais dans les siecles passés il n'y en a point eu de plus magnifique que celle qu'acheva Charles V dans l'appartement de la Reine à l'Hotel St Pol. Depuis le lambris jusques dans la voute, étoit representé sur un fond vert, & dessus une longue terrasse qui regnoit tout autour, une grand forêt pleine d'arbres & d'arbrisseaux, de pommiers, poiriers, cerisiers, pruniers, & autres semblables, chargés de fruits, & entremêlés de lis, de flambes, de roses, & toutes sortes d'atures fleurs: des enfans repandus en plusiers endroits du bois, y cueilloient des fleurs, & mangeoient des fruits: les autres poussoient leurs branches jusques dans la voute peinte de blanc & azur, pour figurer le ciel & le jour; & enfin le tout étoit de beau vert – gai, fait d'orpin & de florée fine." Henri Sauval, *Histoire et recherches des antiquités de la ville de Paris* (Paris: Charles Moette and Jacques Chardon, 1724), 2:281.

22 Ferguson O'Meara, *Monarchy and Consent*, 249–51.

23 Laclotte, *L'École d'Avignon*, 151. For the documents that describe the "chace," see Delachenal, *Histoire de Charles V*, II:375. For the influence of Avignon, Ferguson O'Meara, *Monarchy and Consent*, 244–49.

24 Sauval, *Histoire et recherces*, 2:21.

25 Blume, "Die imaginierte Natur."

26 Blume, "Imaginierte Natur," 472, citing Francesco Petrarca, *Les remèdes aux deux fortunes/De remediis utriusque fortune*, 1354–1366, trans. by Christophe Carraud (Grenoble: Éditions Jérôme Millon, 2002), 304, "enitimini non volucres captare, sed volucres fieri."

27 Blume, "Imaginierte Natur," 473.

28 Roques, "La peintre de la chambre de Clément VI au palais d'Avignon."

29 *Paris 1400*, 225–28.

30 See Woolley, *Medieval Life*.

31 Woolley, *Medieval Life*, 40.

32 Roques, "Peintre de la chambre de Clément VI," 84–85; Dehaisnes, "La tapisserie de haute lisse à arras avant le XVe siècle," 135–36.

33 Krinsky, "Turin-Milan Hours," with full bibliography in note 1. Key recent treatments of the history of the manuscript also include van Buren, Marrow, and Pettenati, *Heures de Turin-Milan*; König, *Die Très Belles Heures von Jean de France, Duc de Berry*; Kemperdick and Lammertse, *The Road to Van Eyck*, 16–19, 98–102, 284–89.

34 Van Buren, Marrow, and Pettenati, *Heures de Turin-Milan*, 342–44.

35 Esther 1:4 (DV).

36 Esther 1:6 (DV).

37 Wilson and Wilson, *A Medieval Mirror*.

38 Krinsky, "Turin-Milan Hours"; Van Buren, Marrow, and Pettenati, *Heures de Turin-Milan*, 344.

39 Franke, *Assuerus und Esther*.

40 For this work and related tapestries, Adelson, *European Tapestry in the Minneapolis Institute of Arts*, 36–51.

41 Deuchler, *Tausendblumenteppich aus der Burgunderbeute*, 22–23.

42 For this work's source documents and analysis, see Deuchler, *Tausendblumenteppich*; Rapp Buri and Stucky-Schürer, *Burgundische Tapisserien*, 115–43.

43 Bonenfant, Fourny, and Le Bon, "Taphonomie de l'Aula Magna de Bruxelles"; Marti, Borchert, and Keck, *Splendour*, 178–79.
44 Zurich Schilling, circa 1480–84, Zurich, Zentralbibliothek, Ms. As. p. 121. For the 1473 meeting and the manuscript with illustration, see Marti, Borchert, and Keck, *Splendour*, 264–65, 331.
45 For the château de Germolles, Beck, Beck, and Duceppe-Lamarre, *Aux marches du Palais*, 104–6; Beck, *Vie de cour en Bourgogne à la fin du Moyen Age*; *Art from the Court of Burgundy*, 146–50; Pinette, *Le château de Germolles*.
46 Beck, Beck, and Duceppe-Lamarre, *Aux marches du Palais*, 104–6; Beck, *Vie de cour*, 107–10.
47 Beck, *Vie de cour*,73–81, esp. 79.
48 Beck, *Vie de cour*, 77.
49 Beck, *Vie de cour*, 109.
50 Beck, *Vie de cour*, 120–23; *Art from the Court of Burgundy*, 148–49.
51 Beck, Beck, and Duceppe-Lamarre, *Aux marches du Palais*, 105; Beck, *Vie de our*, 109.
52 Dehaisnes, "Tapisserie de haute lisse," 135–36.
53 Martin, *Dairy Queens*.
54 Beck, *Vie de cour*, 114–18 with documents.
55 Morand, *Claus Sluter: Artist at the Court of Burgundy*; Prochno, *Die Kartause von Champmol: Grablege der burgundischen Herzöge, 1364–1477*; Recht, "La rhétorique formelle de Claus Sluter, sculpteur du Duc de Bourgogne"; Lindquist, *Agency, Visuality and Society at the Chartreuse de Champmol*. For the well, Nash, "*Well of Moses*." For the tomb, Jugie, *The Mourners*.
56 Nash, "*Well of Moses*," parts 1 and 2.
57 Nash, "*Well of Moses*," part 3, 735–41.
58 Nash, "*Well of Moses*," parts 1–3.
59 For garden fountains in the Valois context, Gertsman, "Sensual Delights: Fountains, Fiction, and Feeling."

References

Adelson, Candace. *European Tapestry in the Minneapolis Institute of Arts*. Minneapolis: The Institute, 1994.
Alexander, Jonathan. "*Labeur* and *Paresse*: Ideological Representations of Medieval Peasant Labor." *The Art Bulletin* 72, no. 5 (1990): 436–52.
Art from the Court of Burgundy: The Patronage of Philip the Bold and John the Fearless, 1364–1419. Cleveland, OH: Cleveland Museum of Art, 2004. Exhibition catalog.
Beaune, Colette. *The Birth of an Ideology*. Translated by Susan Ross Huston. Berkeley: University of California Press, 1991.
Beck, Corinne, Patrice Beck, and François Duceppe-Lamarre. "Les parcs et jardins des ducs de Bourgogne au XIVe siècle: Réalités et representations." In *"Aux marches du palais": Qu'est-ce qu'un palais médiéval? Données historiques et archéologiques; Actes du VIIe Congrès international d'Archéologie Médiévale. Le Mans – Mayenne 9–11 septembre 1999*. Caen: Société d'Archéologie Médiévale, 2001.
Beck, Patrice. *Vie de cour en Bourgogne à la fin du Moyen Age*. Saint-Cyr-sur-Loire: Editions Alan Sutton, 2002.
Blume, Dieter. "Die imaginierte Natur des Papstes: Die *Chambre du Cerf* in Avignon." *Zeitschrift für Kunstgeschichte* 82 (2019): 461–75.
Bonenfant, Pierre P., Michel Fourny, and Madeleine Le Bon. "Taphonomie de l'Aula Magna de Bruxelles: Note archéologique." *Annales de la Sociéte royale d'archéologie de Bruxelles* 65 (2002): 215–34.

Bousmanne, Bernard and Thierry Delcourt, eds. *Miniature flamandes, 1404–1482*. Brussels: KBR, 2011.

Buettner, Brigitte. "Past Presents: New Year's Gifts at the Valois Courts, ca. 1400." *The Art Bulletin*, 83, no. 4 (2001): 598–625.

Camille, Michael. "'For Our Devotion and Pleasure': The Sexual Objects of Jean, Duc de Berry." *Art History* 24, no. 2 (2001): 169–94.

Capellanus, Andreas. *Andreae Capellani regii Francorum De amore libri tres*. Copenhagen: Libraria Gadiana, 1892.

Dehaisnes, Chrétien. "La tapisserie de haute lisse à arras avant le XVe siècle." *Réunion des sociétés savantes et des sociétés des beaux-arts des départements* 3 (1879): 125–39.

Delachenal, Roland. *Histoire de Charles V*. 5 vols. Paris: Picard & fils, 1909–31.

Deuchler, Florens. *Tausendblumenteppich aus der Burgunderbeute: Ein Abbild des Friedens*. Zurich: von Oppersdorff, 1984.

Dückers, Rob and Pieter Roelofs. *The Limbourg Brothers: Nijmegen Masters at the French Court, 1400–1416*. Nijmegen: Ludion, 2005. Exhibition catalog.

Ferguson O'Meara, Carra. *Monarchy and Consent: The Coronation Book of Charles V of France*. London: Harvey Miller, 2001.

Franke, Birgit. *Assuerus und Esther am Burgunderhof: Zur Rezeption des Buches Esther in den Niederländen, 1450–1530*. Berlin: Gebr. Mann, 1998.

Gertsman, Elina. "Sensual Delights: Fountains, Fiction, and Feeling." In *Myth and Mystique: Cleveland's Gothic Table Fountain*, edited by Stephen Fliegel and Elina Gertsman, 59–90. Cleveland, OH: Cleveland Museum of Art, 2016.

Guiffrey, Jules. *Inventaires de Jean Duc de Berry, 1401–16*. 2 vols. Paris: Ernest Leroux, 1894.

Jugie, Sophie. *The Mourners: Tomb Sculptures from the Court of Burgundy*. Dallas: FRAME/French/Regional/American Museum Exchange, 2010.

Kay, Sarah. *The Romance of the Rose*. London: Grant & Cutler, 1995.

Kemperdick, Stephan and Frits Lammertse, eds. *The Road to Van Eyck*. Rotterdam: Museum Boijmans van Beuningen, 2013. Exhibition catalog.

Kendrick, Laura. "L'invention de l'opinion paysanne dans la poésie d'Eustache Deschamps." In *Les "dictez vertueulx" d'Eustache Deschamps: Forme poétique et discours engagé à la fin du moyen âge*, edited by Thierry Lassabatère and Miren Lacassagne, 163–82. Paris: Presses de l'université Paris-Sorbonne, 2005.

König, Eberhard. *Die Très Belles Heures von Jean de France, Duc de Berry*. Munich: Hirmer, 1998.

Krinsky, Carol. "The Turin-Milan Hours: Revised Dating and Attribution." *Journal of Historians of Netherlandish Art* 6, no. 2 (Summer 2014). https://doi.org/10.5092/jhna.2014.6.2.1.

Laclotte, Michel. *L'École d'Avignon*. Paris: Flammarion, 1983.

Lindquist, Sherry. *Agency, Visuality and Society at the Chartreuse de Champmol*. Aldershot: Ashgate, 2008.

Marti, Susan, Till-Holger Borchert, and Gabrielle Keck. *Splendour of the Burgundian Court: Charles the Bold, 1433–1477*. Brussels: Mercatorfonds, 2009. Exhibition catalog.

Martin, Meredith. *Dairy Queens: The Politics of Pastoral Architecture from Catherine de' Medici to Marie-Antoinette*. Cambridge, MA: Harvard University Press, 2011.

McKendrick, Scot. "The Great History of Troy: A Reassessment of the Development of a Secular Theme in Late Medieval Art." *Journal of the Warburg and Courtauld Institutes* 54. (1991), 43–82.

Meiss, Millard. *French Painting in the Time of Jean de Berry: The Late Fourteenth Century and the Patronage of the Duke*, 2nd ed. 2 vols. London: Phaidon, 1969.

Meiss, Millard. *French Painting in the Time of Jean de Berry: The Limbourgs and Their Contemporaries*. 2 vols. New York: G. Braziller, 1974.

Morand, Kathleen. *Claus Sluter: Artist at the Court of Burgundy*. Austin: University of Texas Press, 1991.

Nash, Susie. "Claus Sluter's *Well of Moses* for the Chartreuse de Champmol Reconsidered." Parts 1–3. *The Burlington Magazine*, 147, no. 1233 (2005): 798–809; 148, no. 1240 (2006): 456–67; 150, no. 1268 (2008), 724–41.

Paris 1400: Les arts sous Charles VI. Paris: Fayard and Réunion des musées nationaux, 2004. Exhibition catalog.

Pearson, Andrea. *Gardens of Love and the Limits of Morality in Early Netherlandish Painting*. Leiden: Brill, 2019.

Petrarca, Francesco. *Les remèdes aux deux fortunes/De remediis utriusque fortune*, 1354–1366. Translated by Christophe Carraud. Grenoble: Éditions Jérôme Millon, 2002.

Pinette, Matthieu. *Le château of Germolles*. Germolles: Château of Germolles, 2015.

Plagnieux, Philippe. "La tour Jean Sans Peur: Une épave de la résidence parisienne des ducs de Bourgogne." *Histoire de l'Art* 1–2 (June 1988): 11–20.

Prochno, Renate. *Die Kartause von Champmol: Grablege der burgundischen Herzöge, 1364–1477*. Berlin: Akademie Verlag, 2002.

Rapp Buri, Anna and Monica Stucky-Schürer. *Burgundische Tapisserien*. Munich: Hirmer, 2001.

Recht, Roland. "La rhétorique formelle de Claus Sluter, sculpteur du Duc de Bourgogne." In *Das Porträt vor der Erfindung des Porträts*, edited by Martin Büchsel and Peter Schmidt, 205–17. Mainz: P. von Zabern, 2003.

Roques, Marguerite. "La peintre de la chambre de Clément VI au palais d'Avignon." *Bulletin Monumental* 188, no. 4 (1960): 273–96.

Sauval, Henri. *Histoire et recherches des antiquités de la ville de Paris*. 3 vols. Paris: Charles Moette and Jacques Chardon, 1724.

Schnerb-Lièvre, Marion. *Le Songe du vergier*. Vol. 1. Paris: Editions du Centre national de la recherche scientifique, 1982.

Smith, Geri. *The Medieval French Pastourelle Tradition: Poetic Motivations and Generic Transformations*. Gainesville: University of Florida Press, 2009.

Truitt, Elly. *Medieval Robots: Mechanism, Magic, Nature, and Art*. Philadelphia: University of Pennsylvania Press, 2015.

van Buren, Anne H. *Illuminating Fashion: Dress in the Art of Medieval France and the Netherlands, 1325–1515*. New York: The Morgan Library & Museum, 2011. Exhibition catalog.

van Buren, Anne H., James Marrow, and Silvana Pettenati. *Heures de Turin-Milan, Inv. No. 47, Museo Civico d'Arte Antica, Torino*. 2 vols. Lucerne: Faksimile Verlag, 1996.

Viré, Marc and Agnès Lavoye, "Matériaux et phases de construction: L'étude d'une partie de l'hôtel d'Artois à Paris." In *Pierres du patrimoine européen: Économie de la pierre de l'Antiquité à la fin des temps modernes. Actes du colloque international, Château-Thierry, 18–21 octobre 2005*, ed. by François Blary, Jean-Pierre Gély, and Jacqueline Lorenz, 185–193. Paris: Éditions du Comité des travaux historiques et scientifiques, 2008.

Whiteley, Mary. "'La Grande Vis': Its Development in France from the Mid-Fourteenth to the Mid-Fifteenth Centuries." In *L'Escalier dans l'architecture de la Renaissance: Actes du colloque tenu à Tours du 22 au 26 mai 1979*, 15–20, 219–222. Paris: Picard, 1985.

Whiteley, Mary. "Le Louvre de Charles V: Dispositions et fonctions d'une residence royale," *Revue de l'Art* 97, no. 3 (1992): 60–71.

Whiteley, Mary. "Le Château du Louvre" and "L'hôtel Saint-Pol." In *Paris et Charles V*, edited by Frédéric Pleybert, 113–24. Paris: Action Artistique de la Ville de Paris, 2001.

Wilson, Adrian and Joyce Lancaster Wilson. *A Medieval Mirror*. Berkeley: University of California Press, 1984. http://ark.cdlib.org/ark:/13030/ft7v19p1w6.

Woolley, Linda. *Medieval Life and Leisure in the Devonshire Tapestries*. London: Victoria & Albert Museum, 2007.

4 Garden Theory, Gardening Practice

William and Dorothy Wordsworth

Judith W. Page

According to John Dixon Hunt, in *The Afterlife of Gardens* (2004), garden theories and most gardening practice involve planning the garden and its features according to one's aesthetic assumptions. But when we consider the afterlife of gardens, we add a focus on the responses of future visitors to the garden.[1] Wordsworth incorporates the concept of afterlife into his thinking about the garden, perhaps because of his affective aesthetics and his interest in stories and human contexts. For Wordsworth, and for his sister and some-time collaborator, Dorothy Wordsworth, what happens in and to the garden—its story or plot—is just as important as its beauty or symmetry. This point is not only evident in the poetry, but also in the letters that outline his garden theory and design plans, as well as his ideas on the relationship between nature and art. Although Wordsworth did not write a treatise on gardening, his poems, letters, and reported comments on gardens form a consistent and revealing discourse—one shared and enriched by Dorothy Wordsworth's insights and practice, as her journals reveal.

In this chapter I distinguish Wordsworth's writing about gardens and gardening from the well-considered work on Wordsworth and landscape and more generally on nature.[2] Although there have been some studies of Wordsworth's gardens, there has not been a specific study of Wordsworth as a garden theorist who in many ways rivals and anticipates some influential modern theories of what gardens mean and how they function aesthetically and culturally.[3] Wordsworth was aware of earlier garden-making traditions, both those of the eighteenth century and those of the ancient Romans; he was also forward thinking on this subject, anticipating the Victorian embrace of the natural garden as well as later theorists and historians such as John Dixon Hunt. Furthermore, Wordsworth's formulations on gardens, like many of his other insights into the natural world, were linked to his sister's observations and their shared experiences.

Wordsworth lived amidst several gardens in his early life, but of primary importance is the one that he and his sister planted at Town End in the Lake District village of Grasmere, at what is now known as Dove Cottage. I have argued elsewhere that Dorothy was the head gardener, so to speak, and that William was her helper. This point acknowledges Dorothy Wordsworth's creative agency. But the garden was also a collaboration between sister and brother as they created their home. The garden at Dove Cottage was an early version of the mixed garden or cottage style: a

DOI: 10.4324/9781003381549-5

garden attached to a modest homestead that included both ornamental flowers and plants and vegetables and herbs.[4] It was a garden close to the surrounding beauty of nature, with many wildflowers transplanted in its enclosure, and with the inhabitants of the homestead spending hours growing food, reading, and dreaming in the garden. It was, in other words, a place for labor as well as for leisure and imagination. The Wordsworths thought of their garden as linked to the larger environment of woods and mountains and, therefore, as their most direct ecological connection to the natural world. In many ways this garden set the pattern for the Wordsworths' intertwined thinking about gardens at home and abroad, and about gardens of the future.

When William and Dorothy Wordsworth moved to the cottage on Christmas Eve, 1799, William wrote a letter to Samuel Taylor Coleridge containing the following description:

> D is much pleased with the house and *appurtenances* the orchard especially; in imagination she has already built a seat with a summer shed on the highest platform in this our little domestic slip of mountain. The spot commands a view over the roof of our house, of the lake, the church, helm cragg [sic], and two thirds of the vale. We mean also to enclose the two or three yards of ground between us and the road, this for the sake of a few flowers, and because it will make it more our own. Besides, am I fanciful when I would extend the obligation of gratitude to insensate things?[5]

A few years later Wordsworth will articulate a theory of gardening and place it in the context of other arts. In this letter, he anticipates some of his preoccupations but not in a systematic way. First and foremost, he sees the future garden in the context of home-making—the small, enclosed garden that he and Dorothy see as making the rental property theirs. He credits Dorothy's power of imagination to envision its future. Their garden is a "little domestic slip of mountain," the diminutive serving as an endorsement of the homeliness of their refuge from the turmoil of the world. The mountains, represented by Helm Crag, and the fell visible from Grasmere form a backdrop to the domestic drama that will unfold. The Wordsworths will "enclose the two or three yards of ground" and settle in the cottage bounded by the road, backed by the orchard, and then cradled by the gently rising ground beyond. Wordsworth understands the power of the diminutive, the possibility of a small space managed well and yielding much in reality and in imagination. The poet's rhetorical question about his "gratitude to insensate things"—the objects of nature—reveals that he feels a deep ecological connection to the world of things *as things* but also as spurs to the imagination and anchors of human affection. His gratitude here seems almost a spiritual quality, and in fact the natural world (including the constructed and organized microcosm of the garden) functions as a surrogate religion on this Christmas Eve for the poet who will find "a blessing" in a "gentle breeze" in the opening lines of his epic poem, *The Prelude*.

Although presented in the context of a newsy letter, there are some key concepts here that reveal Wordsworth's understanding of gardens. The garden he and his sister imagine may be an appurtenance of the small cottage, but it will provide a connection to the land that will make the space a place, a home. Throughout his career, Wordsworth links gardens to the comfort, solace, and creativity of home. He, like Dorothy, is less interested in large public gardens or grand estates. As we will see, he also thinks in terms of the topography of the place, of how the garden is situated in the larger landscape and setting: he argues that these elements must be in harmony with each other. The garden at Town End will exist in harmony with the mountains (like Helm Crag, which looms over Grasmere) and lakes, and will be filled with many domesticated wildflowers and mosses collected on long walks, mostly by Dorothy. Far from being a deterrent to imagination, smallness and bounded spaces often paradoxically encourage an openness of thought. And finally, and perhaps as a concept that links all of the above, Wordsworth sees the garden as a place that inspires and deepens human connection via the affections.

It is also striking that Wordsworth imagines the new garden home coming alive, as he imagines its possibilities during the coldest and darkest time of year. He has faith in the seasonal process and rebirth that will allow those "few flowers" to bloom when winter is over. By including the idea of seasonal change in his thinking about gardens, Wordsworth highlights gardens' existence in both space and time. The spatiality of the garden must seem obvious, but the temporality of gardens was an important dimension to both Wordsworths. As Dorothy makes so clear in her Grasmere journals, seasonal change in the garden provides endless delight.

But gardens not only change with the seasons; they also change with the years, sometimes for the better, sometime not. In concluding his letter to his friend and patron Lady Beaumont on his ideas for her winter garden at Coleorton in December 1806—Lady Beaumont and her husband Sir George were building a new home in Lincolnshire—Wordsworth confesses with false modesty that

> I am sensible that I have written a very pretty Romance in this Letter, and when I look at the ground in its present state and think of what it must continue to be, for some years, I am afraid you will call me an Enthusiast and a Visionary.[6]

But it is just this religious zeal and visionary quality that Wordsworth values because it allows him to imagine a future fifty years hence when the garden will be a paradise, his most hopeful version of its afterlife. He imagines this paradisal garden when (likely) neither he nor Lady Beaumont will be there to enjoy it. So, Wordsworth looks forward to the afterlife of the garden, but he also embraces his enthusiastic commitment to build a garden on a site that needs redemption. In the words of a later critic: "The plot of ground where stone had once been quarried, was hollow in shape, something over an acre in size, and about 250 yards from the manor house."[7] The challenge, then, was to create a garden in an inhospitable, abandoned stone quarry.[8]

With the help of workers and a head gardener only identified as "Mr. Craig," as well as frequent consultations with Dorothy, William's plan became a reality, much as he outlined in the sketch for Lady Beaumont.[9] But the afterlife was not so fortunate. Wordsworth will be disappointed in 1841 when he writes to his confidant and dear friend Isabella Fenwick on 24 July about a visit to Coleorton—he does not like the way the garden looks, with an aviary and other features he deplores. His understated conclusion is that it "has not been treated altogether as I wished."[10] Imagine his response to the state of that garden in the twenty-first century, with just traces of its former life peeking through a car park and other "improvements" that Wordsworth could not have imagined. As reported by Peter Dale and Brandon Yen in 2018: "Though a few residual features still commemorate Wordsworth's close personal and creative connections with Coleorton, the main elements of his Winter Garden—not least the cloistral Alley—are long gone."[11] Although we can recover the plot of this garden and its significance through letters, plans, and poems, it is otherwise lost to time. It is not now the kind of romantic ruin that inspired the poet.

Wordsworth, however, provides us with ample narratives of ruins and other lost structures that do inspire. In the untitled poem written in 1797 now called "The Ruined Cottage," Wordsworth's "ruined cottage" is also a ruined garden, with nature reclaiming both the building "four naked walls" (31) and the modest garden:

.... It was a plot
Of garden ground, now wild, its matted weeds
Marked with the steps of those whom as they passed,
The goose-berry trees that shot in long lank slips,
Of currants hanging from their leafless stems
In scanty strings, had tempted to o'erleap
The broken wall.

(54–60)[12]

The narrator tells the story of the garden and how it and the cottage had become a ruin. The "plot/Of garden ground" becomes the plot of the poem and a kind of metaphor for all that Margaret, the young woman on whom the poet meditates, has lost through desertion, war, and poverty. The cottage and garden are not merely picturesque ruins for the casual tourist, but profound, almost sacred, sites that commemorate human suffering; they have a powerful, sympathetic effect on the viewer and, by extension, on those who hear the tale.

Furthermore, in "The Massy Ways, Carried Across These Heights," the poem Wordsworth composed in 1826 when he thought he might have to leave his beloved home and garden at Rydal Mount, the speaker imagines his garden path as perhaps suffering the fate of ancient Roman ways that in Cumbria are no more, at least in their original form. The speaker likens the intellectual and poetic riches he has taken from the garden to "Choice flowers" (17) that are gathered from garden beds; his garden is a kind of palimpsest that bears imaginative traces of its Roman past, which adds depth to his sense of the garden's past and future.

The massy Ways, carried across these heights
By Roman perseverance, are destroyed,

Or hidden under ground, like sleeping worms.
How venture then to hope that Time will spare
This humble Walk?

(1–5)[13]

The poet thus thinks of the garden in terms of its ancient past and uncertain future, hoping for its afterlife, but worried about its fate. But he does not leave it there: he imagines, in a typical move for Wordsworth, that there will be sympathetic "pure Minds" who will preserve his garden path—if not in its material form then in their memories and stories because they "reverence the Muse." By setting the poem in the context of traces of Roman Britain, marked both by persistence and ruin, Wordsworth emphasizes the precariousness of his "humble Walk" in the context of time and change. He would be happy, no doubt, to learn that his beloved pathways and gardens at Rydal Mount yet survive—in the actual place maintained by "pure Minds" and in the minds of those who "reverence the Muse" and read poems about the gardens.

In imagining a time when he might not be able to walk on his path, the speaker highlights (referring to himself in third person) what he most values about the garden, its affective qualities, both in solitary moments and in shared experiences:

No longer, scattering to the heedless winds
The vocal raptures of fresh poesy,
Shall he frequent these precincts; locked no more
In earnest converse with belovéd Friends,
Here will he gather stores of ready bliss,
As from the beds and borders of a garden
Choice flowers are gathered.

(11–17)

Here the poet invokes presence by absence, beginning the passage with the negative "No longer," and then enunciating the very things that he will miss. He also compares the acquisition of memories ("stores of ready bliss") with the ingathering from the actual garden, once again thinking of the garden as a metaphor for creativity and as a source of affections. In the 1835 edition of the poem, Wordsworth identified it as an inscription "Intended to be placed on the door of the further Gravel Terrace if we had quitted Rydal Mount," hence, providing a way of controlling the afterlife of the garden and the response of visitors.[14] Still, as late at 1843, he was worried about the fate of his garden in after-times, musing to Isabella Fenwick that

> I often ask myself what will become of Rydal Mount after our day–will the old walls & steps remain in front of the house & about the grounds, or will they be swept away with all the beautiful mosses & Ferns & Wild Geraniums other flowers which their rude construction suffered & encouraged to grow among them?[15]

He need not have worried: the garden, unlike the winter garden he designed at Coleorton, yet lives.

So concerned was Wordsworth with the affective qualities of the garden that when he begins his disquisition to Sir George Beaumont in his letter of 17 and 24 October 1805, he focuses not on design principles but on the affections and feelings associated with the space—with the way that space become a place linked to the affections:

> Laying out of grounds, as it is called, may be considered as a liberal art, in some sort like Poetry and Painting; and its object like that of all liberal arts is, or ought to be, to move the affections under the control of good sense; that is, those of the best and wisest; but speaking with more precision it is to assist Nature in moving the affections, and . . . the affections of those who have the deepest perception of the beauty of Nature, who have the most valuable feelings, that is, the most permanent, the most independent, the most ennobling, connected with nature and human life.[16]

In imagining the garden at Coleorton, the occasion for this long letter, Wordsworth thus begins with his claim that a garden can function like a work of art because "the laying out of grounds" is in fact a liberal art—and by extension the garden designer is an artist. Like its sister art poetry, garden artistry at its best moves the affections and creates a home for the imagination.[17]

In claiming this elevated status for gardening, Wordsworth goes on to distinguish the medium of garden art: whereas the poet works with words and the painter "colours," the gardener is

> in the midst of the reality of things; of the beauty and harmony, of the joy and happiness of living creatures; of man and children, of birds and beasts, of hills and streams, trees and flowers; with the changes of night and day, evening and morning, summer and winter; and all their unwearied actions and energies, as benign in the spirit that animates them as they are beautiful and grand in that form and clothing which is given to them for the delight of our senses.[18]

The gardener has the particular obligation to use the "reality of things" for delight, aware of the challenge of constant change in any garden. Unlike a poem or painting, which can maintain its form through the years (notwithstanding various textual issues and revisions or new applications of paint), the garden is by definition a place of natural change, a place "connected with nature and human life." The garden, in other words, is not just a place of visual delight (as indicated in the clothing metaphor) but a place that embodies diurnal and seasonal change, a sense of the passage of time and the cycles of life. For the garden, the passage of time is built into the system of meaning-making, with change inevitable. The garden thus provides a poignant reminder of our humanity, of both mortality and the persistence of memory.

Wordsworth the letter writer is akin to Wordsworth the writer of his manifesto, the Preface to *Lyrical Ballads* (1800), who focuses on "the essential passions of the heart" and argues (with a gardening metaphor) that rustic life provides the best soil for nurturing these passions in art. Likewise, Wordsworth advocates simplicity in gardens as being the most effective means of inspiring the garden visitor, of moving the affections toward ennobling thoughts. Wordsworth scorns poetic diction in the Preface where he eschews ornamentation and artificiality—"false refinement or arbitrary innovation"—as impediments to the affective response.[19] The poet decries "gross and violent stimulants" in literature and the arts that inspire momentary thrills rather than deep and lasting emotional connections.[20] Although he does not explicitly include garden art in the Preface, his claim, just a few years later, that gardening is a liberal art allows us to make this connection between poetic and garden rhetoric.

Furthermore, we see this connection explicitly in one of Wordsworth's early sonnets, where the poet links poetic form to gardening concepts. In "Nuns Fret Not" (1802), Wordsworth refers to the sonnet as a "scanty plot of ground," clearly thinking that a small and constrained poem, like a small garden, can produce wonders. Small spaces also provide the psychological discipline for larger creativity, as when the sonnet's boundaries are freeing:

In sundry moods, 'twas pastime to be bound
Within the Sonnet's scanty plot of ground:
Pleased if some Souls (for such there needs must be)
Who have felt the weight of too much liberty,
Should find short solace there, as I have found.
(10–14)

Written at a tumultuous point in his life, this sonnet introduces the multiple layers of "plot," including the political implications of the tug between liberty and restraint.[21] Wordsworth also plays on the meanings of "plot of ground" and his narrative plot, the garden as a place dealing with the material things that require attention and care, as well as an idea or concept for ordering the world. As Robert Pogue Harrison suggests, "like a story, a garden has its own developing plot, as it were, whose intrigues keep the caretaker under more or less constant pressure."[22] Wordsworth frequently thinks of gardens and poems, their form, their creative bounds, and their legacies, interchangeably as art—art using different media but with parallel functions. Even visitors to Wordsworth's long-time home at Rydal Mount (from 1813 to his death in 1850), noted the literary qualities of his garden, as Nathaniel Hawthorne did in 1855 when he remarked that Wordsworth's poetry "had manifested itself in flowers, shrubbery, and ivy."[23]

If we extend this analogy between the rhetoric of poetry and of the garden, elaborate statuary, such as William and Dorothy decried when they visited Isola Bella in Lake Maggiore, functions in garden rhetoric as a kind of poetic diction or false language. Dorothy's comments are parallel to and illuminate William's garden thoughts. According to Dorothy in her *Tour of the Continent* (1820):

> Upon this small island, terraces are crowded together one above another, hewn out of the rock or piled upon that foundation—marble walls, parapets, steps, statues, lemon trees. Here is a Grotto fountain, a pile of stone and marble, high up-raised with monstrous heads intended to spout out water; but not a drop to trickle when we were at the island. Surmounting all, appeared a winged Pegasus, intent upon the sky!—and, to complete the absurdity—below this fountain guarded by monsters, lay a formal garden—not of flowers, but of *stones*, framed out with box-wood into pincushions, half-moons, and like devices, where various-coloured pebbles studiously arranged supplied the place of flowers.[24]

William, I believe, would agree that in this famed garden the balance of art and nature is violated. Dorothy's description (peppered with exclamation marks and other forms of orthographic flourish) represents the garden's equivalent to the false literary style of poetic diction, which William had described in his Appendix on "Poetic Diction" (1802) as "thrust[ing] out of sight the plain humanities of nature

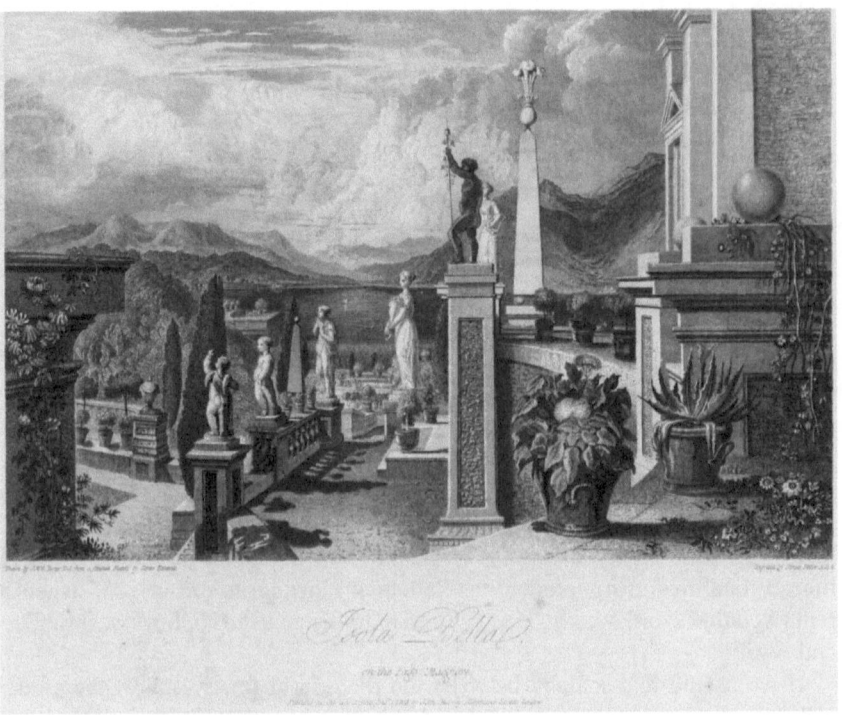

Figure 4.1 Isola Bella, on the Lago Maggiore, 1818. Joseph Mallard William Turner. Engraved by J. Fittler. Transferred from the British Museum 1988. Photo: Tate.

by motely masquerade of tricks, quaintnesses, hieroglyphics, and enigmas."[25] The Isola Bella garden's equivalence of this verbal masquerade are the monsters of rock and stone. We can imagine some of the qualities that the Wordsworths saw in the engraving from Turner of Isola Bella, a space crowded with statuary and constructed objects rather than flowers and plants. As in Dorothy Wordsworth's descriptions, Turner foregrounds a jumble of statuary rather than the natural beauty of a garden, although we do see the mountains in the background as well as striking cloud formations.

Dorothy's own work as a gardener at Dove Cottage attests to a very different aesthetic. The creation of that garden also coincides with William's effort to revise the *Lyrical Ballads* and develop his manifesto. To borrow William's analysis of his own work, Dorothy creates the taste by which her garden can be enjoyed and shuns the analogues of traditional poetic diction in the form of statuary or elaborate structures, in favor of fresh combinations of plants and benches and arbors for the humans who seek solace or inspiration in the garden. Like her brother, Dorothy prefers the "plain humanities of nature" to the "masquerade," finding a sacred blessing in the simple objects of nature. Rather than allusions to classical mythology and or statuary, Dorothy imagines the landscape itself as alive and emotionally expressive. Frequently personifying the flowers in her garden and in the wild, Dorothy suggests a kind of natural animation, as in her description of the daffodils that she and William come upon in Gowbarrow Park on 15 April 1802: "some rested their heads upon [the] stones as on a pillow for weariness and the rest tossed and reeled and danced and seemed as if they verily laughed with the wind that blew upon them over the lake."[26] She brings that same animated language of nature back to the small garden that she cultivates.

In her travel writing (such as the *Tour of the Continent*) that followed the early years in Grasmere, Dorothy stayed true to the aesthetic that both she and William valued. Although, for instance, both she and William appreciated the natural beauty of much of Europe, she could not understand the construction of gardens that do not provide comfort, shelter, and a variety of plants. Her comments echo Wordsworth's written record. In her *Tour of the Continent* Dorothy judges Isola Bella a "*mis*-named" place. Dorothy speculates that the island may have had greater attractions if left as a naked rock rather than "'edified and adorned' by the Italian Dukes,"[27] a reference to the seventeenth-century Baroque palace and elaborate terraced gardens that take up most of the island. Dorothy cannot sanction a garden made primarily of ornamentation and other hard features. Although Dorothy does praise a few flowers and a bit of shade-giving shrubbery as the description develops, her main criticism stands and is intensified by her emphasis on the heat and glare generated by the elaborate stonework. Because she cannot imagine inhabiting such a space, Dorothy does not see it as a garden, although it would fit John Dixon Hunt's definition in *Greater Perfections* of an enclosed and concentrated space, with the stone garden seemingly well situated on the rocky island.[28] For Dorothy there is too much artificiality and ingenuity, without the soft corners for repose and meditation—no green thoughts, no green shade. She objects less to the formality of the garden than to the particular style, which includes mythological allusions and statuary "monsters" that attract attention to themselves. Dorothy Wordsworth, I think, would agree with the anonymous German traveler whom William Hazlitt quotes in his *Notes of a Journey through France and Italy* (1826): the traveler had described this garden as "'a pyramid of sweetmeats,' "[29] an extravagant jumble.

Had she extended her comment into an argument, Dorothy might have invoked reasons similar to William's dislike of poetic diction (akin to fancy ornamentation and figures like the winged Pegasus) as a gimmicky substitute for real emotion. Dorothy objects to the "language" of this garden—it irritates rather than moves her in its unnaturalness. All emblem and no emotion, Isola Bella, spectacular as it was, fails to engage her affections. Twenty years earlier, Hester Piozzi had responded to "a kind of fairy habitation, so like something one has seen represented on theatres," but Dorothy Wordsworth sees this theatricality as too staged and unnatural: the Borromeo gardens on Isola Bella attempt to impose a mythological narrative on the visitor.[30] Dorothy is offended by the enforced mythological grandeur as well as by the human cost of such grandeur: "what a waste of wealth and labour!"[31] For both Dorothy and William, the language of nature is a more trustworthy and powerful stimulant to the imagination. In their view, the designer must work in collaboration with nature, but here nature is upstaged by art.

In contrast to the response to Isola Bella, Dorothy and William had enjoyed the garden on Isola Madre earlier in the same day: here she finds "rich flowers growing in profusion—and fruits of various kinds," although she prefers even there the "wild shrubs and flowers" along the rocky shore. Her valuing of nature, wildness, and the emotional effect of the garden endorse the views that William espouses throughout his career. Her favorite spot on Isola Madre is "at the top of the island" where "there is much of *natural* wildness—trees of various kinds intermingled with grass, vineyards, and beds of flowers, a charming spot for a summer's day's wandering and rest!" Unlike the stone and statuary of Isola Bella, Dorothy had found and praised here the variety of the garden style, with enough diversity to promote extensive wildlife: pheasants, ring-doves, and "a large collection of other birds."[32] Like William, Dorothy wants to see the garden as a microcosm of the natural world, with flora and fauna in the picture. Likewise, when William advises Lady Beaumont about her flower garden in his letter of December, 1806, he suggests that the aesthetic principle of variety within the unity of the garden is essential:

> Thus laid out the Winter Garden would want no variety of colouring, beyond what the flowers and blossoms of many of the shrubs, such as mezereon and laurustinus, and the scarlet berries of the evergreen trees and the various shades of green in their foliage would give.[33]

This aesthetic echoes his claim in the Preface that "a principle which must be well known to those who have made any of the Arts the object of accurate reflection; I mean the pleasure which the mind derives from the perception of similitude in dissimilitude."[34]

William Wordsworth's bedrock concept in his poetics is that art works hand in hand with nature, theory with practice. As he says in the *Guide to the Lakes*, "The rule is simple; with respect to grounds—work, where you can, in the spirit of Nature, with an invisible hand of art."[35] In other words Wordsworth knows that the garden is a constructed space, dependent on the taste and artistry of the designer, but he values art that does not announce itself as such: a humble garden bench as a focal point, say, rather than a winged Pegasus. Because he is so concerned with the respect for nature and setting, Wordsworth repeatedly posits that the house and garden must exist in harmony with the surroundings. In *Guide to the Lakes* Wordsworth rails against whitewashed mansions on mountaintops because they distort the country and disrespect the antiquity of nature and building.[36] His views are not far from those expressed by Austen's narrator in *Pride and Prejudice*, who notes of Pemberley that Elizabeth Bennet "had never seen a place for which nature had done more, or where natural beauty had been so little counteracted by an awkward taste"— although he rarely thinks in terms of the grandeur of a Pemberley.[37]

Readers of Wordsworth know that in his poetry—a good example would be "Tintern Abbey"—he creates a fiction of spontaneity that makes the poetry seem effortless: we know that this poem, for instance, did not come into being as a spontaneous overflow without thought and revision following the visit to a height above the ruined abbey, but the poem itself has a conversational quality and Wordsworth's language gives that sense of spontaneity:

> The day is come when I again repose
> Here, under this dark sycamore, and view
> These plots of cottage-ground, the orchard tufts,
> Which, at this season, with their unripe fruits,
> Among the woods and copses lose themselves,
> Nor, with their green and simple hue, disturb
> The wild green landscape. Once again, I see
> These hedgerows, hardly hedgerows, little lines
> Of sportive wood run wild . . .
>
> (9–17)

We follow the viewer as he scans the plots of ground and orchards as they disappear into a sea of green, as do the hedgerows, which are revised as "hardly hedgerows" in imitation of the mind that processes the scene. Thus, the revision of thought and landscape takes place before our eyes and ears as we read or listen to the seemingly spontaneous thought, with the flowing enjambments ("view/These plots" or "see/These hedgerows") propelling us forward. But what seems spontaneous is carefully crafted blank verse, containing and ordering the very wildness described in the lines and giving us a way to process and transform nature into art.

Dorothy Wordsworth was also acutely sensitive to the relationship between wild and cultivated nature. When visiting Bothwell Castle in 1803, she records in her journal that

> I was hurt to see that flower-borders had taken the place of the natural overgrowings of the ruin, the scattered stones, and wild plants. It is a large and

grand pile of red freestone, harmonizing perfectly with the rocks of the river, from which, no doubt, it had been hewn. When I was little accustomed to the unnaturalness of the modern garden, I could not help admiring the excessive beauty and luxuriance of some of the plants ... which scrambled up the castle walls, along with ivy, and spread its vine-like branches so lavishly that it seemed to be in its natural situation ... though not self-planted among the ruins of this country, it must somewhere have its native abode in such places.[38]

Dorothy's thought processes, which she records as a seemingly spontaneous flow, are as interesting as William's in the passage above: she begins with a statement of her disappointment that a "modern garden" was planted so close to the ruin, but amends her thought with the admission that "I could not help" admiring the cultivated plants that began to act so wantonly, climbing up the walls of the castle, that they seemed to be "natural" and wild. She concludes by welcoming the plants, which may not be native to the spot but have their "native abode" somewhere akin to this scene. She thus reaches a generous conclusion that artful plantings can appear natural—or that cultivated plants can become natives when they live in harmony with their surroundings.

Similarly, Wordsworth's gardening aesthetic often hides the fact of its artistry. For instance, like all good designers, Wordsworth recognizes that boundaries and enclosure are essential to a garden, but he seems to value most those boundaries that appear most natural. As we will see, he likes boundaries formed by plantings themselves and if there is a wall, let it be claimed by nature. For instance, in his *Guide to the Lakes*, Wordsworth mocks a neighbor who criticized his garden wall:

'If I had to do with this garden . . . I would sweep away all the black and dirty stuff from that wall.' The wall was backed by a bank of earth, and was exquisitely decorated with ivy, flowers, moss, and ferns, such as grow of themselves in like places; but the mere notion of fitness associated with a trim garden-wall prevented, in this instance, all sense of the spontaneous bounty and delicate care of nature.[39]

Wordsworth values precisely those qualities that his neighbor, "a most respectable person," denigrates: the signs of natural processes and the passage of time, which allows a constructed object (a wall) to become akin to a natural object. The gardener anticipates the "spontaneous bounty" of nature, which becomes a part of his art.

Hence, in writing to Lady Beaumont about the boundary line of evergreens at Coleorton, Wordsworth is careful to specify the need for this demarcation but also aware that this boundary will consist of living things—not a new wall constructed around the planned winter garden. Likewise, he later writes a poem, "A Flower Garden" (composed in 1824 and published in 1827), celebrating the summer flowers that Lady Beaumont has planted in "connection" with Coleorton. In doing so, she has fulfilled his advice that although the place was "consecrated" as a winter garden it could be beautiful in all seasons—and this notion brings in the element of time. Wordsworth praises the boundary of the garden because it has become such a natural part of the place.[40]

> Yet, where the guardian fence is wound,
> So subtly are our eyes beguiled,
> We see not nor suspect a bound,
> No more than in some forest wild;
> The sight is free as air—or crost
> Only by art in nature lost.
> . . ..
> Apt emblem (for reproof of pride)
> This delicate Enclosure shows
> Of modest kindness, that would hide
> The firm protection she bestows;
> Of manner, like its viewless fence,
> Ensuring peace to innocence.[41]

The speaker values the boundary because it is a viewless fence, in other words, unperceived as a fence—as in Keats' "viewless wings of Poesy" in "Ode to a Nightingale."[42] Like the house and garden that merge with nature in *Guide to the Lakes*, the boundary here is all the more wonderful because it is unsuspected. It provides the protection of an enclosed space—a garden—by means of the invisible hand of art. As Hunt argues in *Greater Perfections*, a garden's concentration of effects can form the illusion of enclosure without an actual boundary wall or fence.[43] This concentration, furthermore, links the garden enclosure to lyric poetry, which also depends on the boundary lines of form, in this case an established rhyme scheme and stanzaic pattern. Wordsworth clearly had made this connection in "Nuns Fret Not."

In addition to a concentrated and enclosed space as a defining feature of a garden, Wordsworth values garden pathways, when space permits, because pathways represent possibility and provide the structure for both solitude and sociability. Garden pathways, which invite movement because they lead somewhere, are inherently narrative. Wordsworth states in the letter to Lady Beaumont:

> The Path of which I have been speaking should wind around the garden, mostly near the Boundary line, which would in general be seen or felt, as has been described, but not always; for in some places particularly near the high Road it would be kept out of sight, so that imagination might have room to play. It might perhaps with propriety lead along a second border under the clipped holly hedge; everywhere else it should only be accompanied by wildflowers.[44]

Here Wordsworth implies that pathways are vehicles of the imagination in a garden: they lead the participant on but also add to the element of secrecy, wonder, and surprise that comes with the best gardens. Pathways also provide meditative space for those who think best while walking, but if a path is wide enough it also provides the opportunity for sharing and sociability. The speaker refers to both of these qualities in his "Massy Ways" poem. Furthermore, in his earlier letter to Sir George Beaumont, Wordsworth had praised the path along the river in the Lowther estate because it afforded just such sociability when he came upon some peripatetic musicians. For Wordsworth and many other writers and designers, the garden provides the space for private meditation and creativity as well as for shared experiences—the two are not exclusive. As a vehicle of imagination, the pathway also has a metaphorical and poetic dimension.

Wordsworth had found this poetic element in his reading on gardens. In his letter to Sir George, for instance, Wordsworth alludes to Joseph Addison's *Spectator* (No. 477, Sept. 6, 1712) as a source for his ideas about the winter garden. Addison encourages the use of evergreens and indicates a keen understanding that a well-structured garden in winter is a thing of beauty and not a site of death and decay. He also praises the upper garden at Kensington, which had been a gravel-pit (not unlike the abandoned quarry that Wordsworth encounters), from which "an unsightly Hollow" is transformed "into so beautiful an Area." In addition, Addison offers a particularly literary reading of garden styles, and I imagine this passage had an equally strong influence on Wordsworth: "I think there are so many kinds of Gardening as of Poetry: Your Makers of Parterres and Flower-Gardens are Epigrammatists and Sonneteers in this Art: Contrivers of Bowers and Grottos, Treillages and Cascades, are Romance Writers . . . "[45] Likewise, Wordsworth tells Lady Beaumont that he speaks of a Bower "such as you will find described in the beginning of Chaucer's Poem of The Flower and the Leaf, and also in the beginning of the Assembly of Ladies," connecting his garden art to his English literary ancestry and to two poems that depend on the privacy and seclusion of the garden.[46] In the same letter, Wordsworth refers to his proposed garden as including "a little Parlour of verdure" (echoing the "pretty parlour" of the Flower and Leaf) and a "moss'd seat and small stone table."[47] Wordsworth articulates a garden that consists of compartments or rooms parallel to indoor spaces, a style of garden that defines two of the most remarkable gardens created in the twentieth century in England: Lawrence Johnston's Hidcote Manor and Vita Sackville-West's Sissinghurst Castle Garden.[48] If we go back to Addison, we also find the connection of garden spaces to different literary genres and to the association of stanzas with "little rooms." And like literary genres, different types of gardens provide the framework for different meanings, whether it be the sonnet's scanty plot of ground or the epic's encyclopedic expanse and development of plot.

Although we might look toward Hidcote and Sissinghurst as twentieth-century models of gardens with distinctive rooms, Wordsworth, an astute scholar of Latin literature, may have been thinking of the Romans when he imagined his garden rooms for the winter garden at Coleorton and his connection of garden spaces to poetic genres and rhetoric also may have ancient roots. I am thinking particularly of Pliny the Younger's so-called "villa letters," comprising letters in Books 2 (Letter 17), 5 (Letter 6), and 9 (Letter 7). Scholars know that Wordsworth read these letters, for he refers to them in a letter to Coleridge (April 16, 1802) and the letters are included in the catalogue of Wordworth's library.[49] In the villa letters, Pliny refers to the rooms of his villas as overlapping with the rooms of his gardens. He also sees the collection of rooms in his villas as akin to the literary collection of his letters—different rooms and different letters have different functions in relation to the whole. At first one might think of the distinctions: Pliny describes these rooms as both outer (along an extensive colonnade, with courtyards) and inner (for dining and sleeping in proximity to the outside space). He is describing, after all, an Italian setting and not the cold, mountainous north of England. However, as early as the Grasmere years, the Wordsworths thought of their garden as an extension of their small cottage—even in terms of sections or rooms for outdoor living, as in the summerhouse or shed.

Furthermore, Wordsworth advocated linking the house to its natural environment. As Bettina Bergmann reminds us, Pliny was also interested in setting "certain built spaces in relation to the environment" rather than architectural precision and in this sense his aesthetic foreshadows Wordsworth's much later concern for the relationship between structure and the natural site, building and nature.[50] In describing his Tuscan estate, Pliny writes that

> In front of the colonnade is a terrace laid out with box hedges clipped into different shapes, from which a bank slopes down, also with figures of animals cut out of box facing each other on either side. On the level below there are waves—or I might have said ripples—a bed of acanthus. All round is a path edged by bushes which are trained and —cut into different shapes, and then a drive, oval like a racecourse, inside which are various box figures and clipped dwarf shrubs. The whole garden is enclosed by a dry-stone wall which is hidden from sight by a box hedge planted in tiers; outside is a meadow, as well worth seeing for its natural beauty as the formal garden I have described; then fields and many more meadows and woods.[51]

Although it is hard to visualize this garden completely, we can see that certain qualities—the bordering stone wall hidden by the box hedge, for instance—anticipate Wordsworth's concern with boundaries constructed of natural materials that conceal some of the artistry. The bounding path is reminiscent of Wordsworth's plan for the winter garden, although Wordsworth was not as fond of fanciful topiary as the Romans. It is also striking that Pliny orders his description, moving from the intricacies of the enclosed formal garden with topiary, to the larger surrounding meadows and then the woods—moving from built to wild, as also found in many later British estates.

In Wordsworth's thinking, a garden reflects the character as much as the aesthetics of its creator, a concept central to Pliny.[52] Roy Gibson argues that the villa letters "necessarily offer a portrait of their owner" and reveal "the tendency within elite Roman culture to see the villa—like style of speech—as a reflection of the character of the man." The emphasis is not on extravagance but good taste—there is a rhetoric to the villa, a way of communicating with the world.[53] Wordsworth also conveys these concepts in a series of letters to the Beaumonts. We recall that in the letter to his patron Wordsworth had also been concerned with the rhetoric of grounds; he lectures to Sir George on character and taste, stating that

> I know nothing which to me would be so pleasing or affecting as to be able to say when I am in the midst of a large estate, this Man is not the victim of his Condition; he is not the Spoiled Child of worldly grandeur; the thought of himself does not take the lead in his enjoyments; he is, where he ought to be, lowly minded, and has human feelings; he has a true relish of simplicity . . . let nature be all in all, taking care that every thing done by man shall be in the way of being adopted by her. If people chuse that a great mansion should be the chief figure in a Country, let this kind of keeping prevail through the picture, and true taste will find no fault.[54]

Perhaps with a bit more of an interest in romantic simplicity than faith in the benevolence of the elite, Wordsworth insists that his man of taste be guided by nature and human feeling, like the natural forest path in Lowther Woods (winding "under trees with the wantonness of a River or a living Creature"). In concluding, Wordsworth says that

> if I were disposed to write a sermon, and this is something like one, upon the subject of taste in natural beauty I should take for my text the little pathway in Lowther Woods, and all which I had to say would begin and end in the human heart.[55]

In other words, Wordsworth says that if he were inclined to write a theory of gardens ("a sermon"!), he would start with an actual place (in this case the "little pathway") linked to the affections. In planning his garden, Wordsworth insists to Beaumont that these affective qualities take the lead.

If Wordsworth looks back to Roman models and to the previous century, his theories of the garden and of natural beauty also propel us forward. Victorian gardeners, with whom Wordsworth, who lived from 1770–1850, overlapped, were attuned to these differences. The mid-Victorian period would see a resurgence of formality—the trim garden, with formal beds of annuals, known as carpet bedding. This style is perhaps best captured by the prolific Victorian garden writer, Shirley Hibberd, who commented that

> Modern taste requires that a garden should be a garden, not a wilderness, nor a prairie, nor a 'boundless contiguity of shade.' It is an artificial affair, or wild weeds would be allowed to riot in it; it is trim and bright, and in some sense a picture, and hence a framing or boundary of some kind is essential. If boundary is essential why should it not be visible, and in its way ornamental.[56]

Hibberd wrote this in his *Rustic Adornments* in 1856, just a few years after Wordsworth had died, and we can be pretty sure that his ideas would not have pleased the poet who preferred "viewless" boundaries and wildflowers to exotic annuals. As we have seen, trimness was not a Wordsworthian value. In fact, in notes to his poems dictated to Isabella Fenwick in 1843, Wordsworth comments on his poem

"Poor Robin" (1840): "Defend us from the tyranny of trimness & neatness . . . Weeds have been called flowers out of place."[57]

In 1870 came William Robinson's *The Natural Garden*, which would sweep away large swathes of bedding plants and artificial styles in favor of just the aesthetic that the Wordsworths had celebrated. For instance, Robinson particularly loved the sight of wild roses weaving their way through trees, a style that Sackville-West, who admired Robinson, later adapted at Sissinghurst. But this style already had Wordsworth's approval. Consider his sonnet "How Sweet It Is" (1807), where the poet describes the "wild rose tip-toe upon hawthorn stocks."[58] And in his letter to Lady Beaumont, he states:

> Few of the more minute rural appearances please me more than these, of one shrub or flower lending its ornaments to another; there is a pretty instance of this kind now to be seen near Mr. Craig's [the gardener's] new walk; a bramble which has furnished a wild Rose with its green leaves, while the Rose in turn with its red lips has to the utmost of its power embellished the Bramble . . .[59]

Likewise, in his letter on the Kendal and Windermere Railway (1835), Wordsworth speaks of a garden wall that the casual tourist or "improver" would want to sweep away in favor of a "trim garden wall," as we saw earlier.[60] The word "trim" that Wordsworth uses here in 1835 would by mid-century become a code for the kind of garden that Wordsworth—and Robinson and later Sackville-West—would scorn, in favor of the natural garden, not purely natural of course but designed with artistry that conceals itself: the invisible hand of art. I would add that this garden aesthetic is in itself collaborative, in that it values the ways that one feature or natural object allows another to thrive in tandem with it—roses depending on brambles or ferns and mosses on the crevasses of old walls. Nature might be guided and coaxed, but not regimented into trimness: the best art collaborates with nature.

The seeds of this natural style perhaps reach back not to Coleorton or Rydal Mount but to the garden at Town End. For, as I have said, here was a true cottage garden, a mingled style in which beans and peas dallied with honeysuckle, transplanted wildflowers with carrots. Although Wordsworth conceived of his later gardens as more expansive—the subjects of much discussion and design—his roots were nonetheless in the most modest plot that he cultivated with his sister and that sustained him as he went on to the more complex and expansive genres. But the story of this garden also uncovers the challenges of planting a garden that one might not be able to control and certainly cannot keep in its spring-time freshness for all time. What will be the garden's afterlife? For many years now, the garden has been lovingly restored and is open to the public at Dove Cottage. But what of its immediate afterlife? Dorothy's Grasmere journal tells of the creation of this garden with impassioned immediacy and William's poem "A Farewell" both commemorates the garden and prefigures its next phase: the garden will remain but will take on new meanings as William marries and begins to raise a young family in that small cottage.[61] But that is another story, deeply intwined with the siblings' biographies and best told outside the boundaries of the garden. Suffice it to say for now that this garden marks the beginning of a long engagement with gardens as real places and as central to the Wordsworths' ideas about art, nature, and the human heart.

Notes

1   See, especially, *The Afterlife of Gardens,* 1–32.
2   The most comprehensive study of Wordsworth and landscape is still Noyes, *Wordsworth and the Art of Landscape,* published over half a century ago.
3   Buchanan provides an introduction to the gardens; see, more recently, Dale and Yen, who do intersperse commentary on Wordsworth's theory, although that is not their main focus.
4   See Page and Smith, especially 151–53 on the cottage garden style in relation to the Grasmere garden.
5   *Letters of William and Dorothy Wordsworth: The Early Years,* 274–75.
6   *Letters of William and Dorothy Wordsworth: The Middle Years,* 119.
7   Noyes, *Wordsworth and the Art of Landscape,* 113.
8   For a more extended description of the topography, see Anderson, "Wordsworth and the Gardens of Coleorton Hall," 210–11.
9   It seems that William's letter of December 1806 was enclosed within Dorothy's letter of December 23, both of them to Lady Beaumont and both now at the Pierpont Morgan Library. The plan is included in Dorothy's letter—in the middle of the letter and (rather mysteriously) drawn upside down. See https://www.themorgan.org/literary-historical/295623. Accessed October 11, 2022.
10 *Letters of William and Dorothy Wordsworth: The Later Years,* 217.
11 Dale and Yen, *Wordsworth's Gardens and Flowers,* 11. Noyes, *Wordsworth and the Art of Landscape,* also describes a few remnants of the garden in 1968, 123.
12 All references to Wordsworth's poetry are to Gill, unless otherwise noted; line references will be given in the text when references are to Gill.
13 *Poetical Works of William Wordsworth,* vol. 4: 201–2.
14 *Poetical Works of William Wordsworth,* vol. 4: 201.
15 *Poetical Works of William Wordsworth,* vol. 4: 437–38.
16 *Letters of William and Dorothy Wordsworth: The Early Years,* 627.
17 In comparison, Thomas Jefferson wrote a letter to his granddaughter the same year in which he also referred to gardening as a fine art. See *Thomas Jefferson's Garden Book,* 303–4. Also cited in Way, 11.
18 *Letters of William and Dorothy Wordsworth: The Early Years,* 627.
19 *Prose Works* 1: 124.
20 *Prose Works* 1: 129.
21 For a political and formal reading of this and other sonnets, see Page, "Wordsworth's French Revolution," *Wordsworth and the Cultivation of Women,* 54–76.
22 Harrison, *Gardens,* 7.
23 Hawthorne, "English Notebooks," 8: 27.
24 *Journals of Dorothy Wordsworth,* 253.
25 *Prose Works* 1: 161–62.
26 *Grasmere Journals,* 85.
27 *Journals of Dorothy Wordsworth,* 252.
28 Hunt, *Greater Perfections,* says that "Gardens are privileged . . . because they are concentrated or perfected forms of place-making" 11.
29 Hazlitt, *Notes of a Journey through France and Italy,* 278. Hazlitt goes on to say "I had supposed this to be a heavy German conceit, but it is a literal description. The pictures in the Palace are trash" (278).
30 Qtd. in Hunt, *Garden and Grove,* 98.
31 *Journals of Dorothy Wordsworth,* 253.
32 *Journals of Dorothy Wordsworth,* 252.
33 *Letters of William and Dorothy Wordsworth: The Middle Years,* 118.
34 *Prose Works* 1: 148.
35 *Guide to the Lakes,* 74.

36 *Guide to the Lakes*, 80–81.
37 *Pride and Prejudice*, 271.
38 Qtd. in *Poetical Works*, vol. 3: 532–33.
39 *Guide to the Lakes*, 151.
40 *Letters of William and Dorothy Wordsworth: The Middle Years*, 118.
41 *Poetical Words of William Wordsworth*, vol. 2: 126–27; lines 25–30 and 43–48.
42 Keats, "Ode to a Nightingale," stanza 4.
43 Hunt, *Greater Perfections*, 20.
44 *Letters of William and Dorothy Wordsworth: The Middle Years*, 115.
45 Addison, *Spectator*, 4: 13.
46 Modern scholarship has assigned these poems to an anonymous "lady" and not Chaucer. See Derek Pearsall's commentary: https://d.lib.rochester.edu/teams/test/pearsall-flour-and-the-leafe-introduction. Accessed Oct. 11, 2022.
47 *Letters of William and Dorothy Wordsworth: The Middle Years*, 117–18.
48 *Wordsworth's Gardens and Flowers*, make the connection, 79. For an analysis of the garden rooms in these two English gardens, see Page and Smith, "Castle and Rose: Vita Sackville-West and the Redemption of Sissinghurst," *Women, Literature, and the Arts of the Countryside in Early Twentieth-Century England*, 216–21.
49 Wordsworth claims in the letter that he read Pliny's letters "years ago." See *The Letters of William and Dorothy Wordsworth: the Early Years*, 347–48. For the listing in Wordsworth's library, see Shaver and Shaver, *Wordsworth's Library*, 202.
50 Bergmann, "Visualizing Pliny's Villas," 409.
51 *Letters of Pliny the Younger*, 5.6 140.
52 Austen taps into the same connection between estate and character in *Pride and Prejudice*, when in volume 3, chapter 1 Elizabeth Bennet visits Pemberley and learns to read the house and grounds in terms of Darcy's character.
53 Gibson, "Reading the Villa Letters," 216–17.
54 *Letters of William and Dorothy Wordsworth: The Early Years*, 625. Wordsworth's language echoes his claim in "Tintern Abbey" that nature was "all in all" (line 76).
55 *Letters of William and Dorothy Wordsworth: The Early Years*, 626, 628.
56 Hibberd, *Rustic Adornments*, 359.
57 *The Poetical Works of William Wordsworth*, vol. 4: 438.
58 *The Poetical Works of William Wordsworth*, vol. 3: 21, line 5.
59 *Letters of William and Dorothy Wordsworth: The Middle Years*, 120.
60 *In Guide to the Lakes*, 151.
61 The poem was written in 1802 and first published in 1815 as "Farewell, Thou Little Nook of Mountain Ground." It was known as "A Farewell" after the 1827 edition of Wordsworth's poetry.

References

Anderson, Anne. "Wordsworth and the Gardens of Coleorton Hall." *Garden History* 22.2 (Winter 1994): 206–17.

Addison, Joseph, Richard Steele, and Others. *The Spectator*. 4 vols. Ed. G. Gregory Smith. London: J. M. Dent & Sons Ltd., 1907.

Austen, Jane. *Pride and Prejudice*. Ed. Pat Rogers. New York: Cambridge University Press, 2013.

Bergmann, Bettina. "Visualizing Pliny's Villas." *The Journal of Roman Archeology* 8 (1995). 406–20.

Buchanan, Carol. *Wordsworth's Gardens*. Lubbock: Texas Tech University Press, 2001.

Dale, Peter and Brandon C. Yen. *Wordsworth's Gardens and Flowers: The Spirit of Paradise*. Woodbridge, Suffolk: ACC Art Books, 2018.

Gibson, Roy K. "Reading the Villa Letters 9.7, 2.17, 5.6," in *Reading the Letters of Pliny the Younger: An Introduction*. Eds. Roy K. Gibson and Ruth Morello. New York: Cambridge University Press, 2012. 200–33.

Gill, Stephen. *William Wordsworth: A Life*. Oxford: Clarendon, 1989.

Harrison, Robert Pogue. *Gardens: An Essay on the Human Condition*. Chicago: The University of Chicago Press, 2008.

Hawthorne, Nathaniel. "English Note-Books." *The Complete Works of Nathaniel Hawthorne*, 22 vols. Boston: Houghton Mifflin, 1899.

Hazlitt, William. *Notes of a Journey through France and Italy* (1826). In *The Collected Works of William Hazlitt*, 12 vols. Eds. A. R. Waller and Arnold Glover. London: J. M. Dent, 1903. Vol. 9: 89–303.

Hibberd, Shirley. *Rustic Adornments for Homes of Taste and Recreations for Town Folk in the Study and Imitation of Nature*. Intro. John Sales. 1856. London: Century Hutchinson, 1987.

Hunt, John Dixon. *Garden and Grove, The Italian Renaissance Garden in the English Imagination: 1600–1750*. Princeton, NJ: Princeton University Press, 1986.

Hunt, John Dixon. "The Garden a Cultural Object." *Denatured Visions: Landscape and Culture in the Twentieth Century*. Eds. Stuart Wrede and William Howard Adams. New York: Museum of Modern Art, 1991. 19–32.

Hunt, John Dixon. *Greater Perfections: The Practice of Garden Theory*. Philadelphia: University of Pennsylvania Press, 2000.

Hunt, John Dixon. *The Afterlife of Gardens*. Philadelphia: University of Pennsylvania Press, 2004.

Jefferson, Thomas. *Thomas Jefferson's Garden Book, 1766–1824. With Relevant Extracts from His Other Writing*. Philadelphia, PA: American Philosophical Society, 1944.

Keats, John. *The Poems of John Keats*. Ed. H. W. Garrod. New York: Oxford University Press, 1966.

Noyes, Russell. *Wordsworth and the Art of Landscape*. Bloomington: University of Indiana Press, 1968.

Page, Judith W. "Wordsworth's French Revolution: The Sonnets of 1802." *Wordsworth and Cultivation of Women*. Berkeley: University of California Press, 1994. 54–76.

Page, Judith W. and Elise L. Smith. "Dorothy Wordsworth: Gardening, Self-fashioning, and the Creation of Home." *Women, Literature, and the Domesticated Landscape: England's Disciples of Flora, 1780–1870*. New York: Cambridge University Press, 2011. 139–62.

Page, Judith W. and Elise L. Smith. "Castle and Rose: Vita Sackville-West and the Redemption of Sissinghurst." *Women, Literature, and the Arts of the Countryside in Early Twentieth-Century England*. New York: Cambridge University Press, 2021. 200–21.

Pliny [Plinius Caecilius Secundus, Caius]. *The Letters of Pliny the Younger*. Trans. Betty Radice. New York: Penguin, 1969.

Shaver, Chester L. and Alice C. Shaver. *Wordsworth's Library: A Catalogue*. New York: Garland Publishing, 1979.

Way, Thaïsa. *Unbounded Practice: Woman and Landscape Architecture in the Early Twentieth Century*. Charlottesville: University of Virginia Press, 2009.

Wordsworth, Dorothy. *Journals of Dorothy Wordsworth*. 2 vols. Ed. Ernest de Selincourt. New York: Macmillan, 1941.

Wordsworth, Dorothy. *The Grasmere Journals*. Ed. Pamela Woof. New York: Oxford University Press, 1991.

Wordsworth, William. *The Poetical Works of William Wordsworth*. Eds. Ernest de Selincourt and Helen Darbishire. 5 vols. Oxford: Clarendon, 1940–49.

Wordsworth, William. *The Prose Works of William Wordsworth*. 3 vols. Eds. W. J. B. Owen and Jane Worthington Smyser. Oxford: Clarendon, 1974.

Wordsworth, William. *Guide to the Lakes*. 5th ed. Ed. Ernest de Selincourt. New York: Oxford University Press, 1977.

Wordsworth, William. *William Wordsworth: The Major Works*. Ed. Stephen Gill. New York: Oxford University Press, 1984.

Wordsworth, William and Dorothy Wordsworth. *The Letters of William and Dorothy Wordsworth: The Early Years, 1787–1805*. 2nd ed. Ed. Ernest de Selincourt. Rev. Chester L. Shaver. Oxford: Clarendon, 1967.

Wordsworth, William and Dorothy Wordsworth. *The Letters of William and Dorothy Wordsworth: The Later Years*. Part 4, 1840–1853. 2nd ed. Ed. Ernest de Selincourt. Rev. Alan G. Hill. Oxford: Clarendon Press, 1988.

Wordsworth, William and Dorothy Wordsworth. *The Letters of William and Dorothy Wordsworth: The Middle Years*. Part 1: 1806–1811. Ed. Ernest de Selincourt. Rev. Mary Moorman. Oxford: Clarendon Press, 1969.

5 Places for the Spirit, Photographs of Traditional African American Gardens

Vaughn Sills

In 1987 I began photographing traditional African American gardens and the women and men who created them. Fifty years before I began my project, the twentieth-century fiction writer Eudora Welty, who lived all her life in Jackson, Mississippi, worked for the WPA photographing African Americans during the Depression. In 1971, in the foreword to her book of photographs, *One Time, One Place*, she wrote:

> In taking all these pictures, I was attended, I now know, by an angel—a presence of trust. In particular, the photographs of black persons by a white person may not testify soon again to such intimacy. It is the trust that dates the pictures now, more than the vanished years.[1]

Many years later, I felt honored to have been given what felt like a similar trust. At the same time, I want to acknowledge that there have always been questions in my mind and from viewers of my work about my being white and photographing a part of African American culture: Was I trying to tell a story that was not mine to tell? Did I have a right to make the images and assert knowledge about the history and cultural significance of the gardens? Those who view the work will, in the end, answer these questions for themselves, in a cultural and political context that will continue to evolve. Here I will respond to them indirectly by sharing what I thought and felt as a white person photographing a part of Black culture. My understanding of this experience is intertwined with the significance of the gardens themselves—the true focus of my work and of this chapter.

When I, a white woman from the North, began this work in 1987, I was forty years old. I lived, as I still do, in Massachusetts and was teaching photography as an adjunct professor at Simmons College (now Simmons University), where I would later become tenured. My cultural heritage is from the maritime provinces of Canada—with English, Irish, and Scottish ancestors; and this place, particularly Prince Edward Island, has been the source of three of my photography projects. Our family moved to the States when I was young, and after several years in northern New Hampshire, in 1956 we drove to Louisiana and, soon after, my family moved to Baton Rouge, Louisiana. My father was an industrial engineer who moved from job to job; my mother was a teacher before I was born, a home-maker, and later

DOI: 10.4324/9781003381549-6

a small bookstore owner. I was nine when I first saw, with both puzzlement and astonishment, signs that said "Colored" and "White," and ten when I started school where petticoats and flats were *de rigueur* for the young girls and was called Yankee (which was an epithet) by my classmates, and I began to learn about race and racism. Considering its significance in American society, I was old to be learning those lessons about being white. As research on racial identity development has shown, a young Black girl would certainly have been conscious of her place in the racial hierarchy well before the age of ten, while I was kept from this necessity by being white. And I'm sure I was slow to catch on, partly because I didn't have to learn and partly because I was most concerned at that point with figuring out how to fit in, how to sound like I belonged (my accent changed and I learned to say "Yes ma'am" to my teachers pretty fast), how to make new friends through new playground games, and a new social etiquette. What I was given to fit into, as a consequence of our segregated neighborhood, was all white.

Even so, it wasn't long before I began to learn about race and economic class—in two ways: First, most prominently, there was talk—constant talk, among children as well as among our teachers—about the integration of schools. Second, unlike the elementary school I'd gone to in New Hampshire, the schools and neighborhoods in Louisiana were also segregated by economic class. We lived in a small development of new middle-class houses, racially segregated due to decades of pernicious policies, with its own elementary and secondary schools; thus my peers and our teachers were all white and middle class. I was almost sixteen when my parents decided to leave the South. But before that I had been quite vocal about my belief that schools should be integrated, and I had been called a good deal more nasty names than "Yankee"— unprintable today—not merely by my peers, but by a teacher. We moved back North, this time to Maine. In my early twenties I had the opportunity to travel to Ghana, Nigeria, the Ivory Coast, and Mali, where I learned about the rich and varied cultures of several African peoples and to see their strong vibrant ways of life—which led me to appreciate and learn more about African art and, in retrospect, to be open to what I would see and learn in the photography project presented here. I have remained committed to working against racism and was active in various ways (whether effective or not) throughout my adult life. But in 1987, despite all these experiences and concern about racism, when I began this project there was still much I did not know.

I was introduced to the gardens in 1987 by a scholar and long-time friend, Sara Glickman. Sara, who is white, had researched and written a master's thesis on African American settlement patterns in and around Athens, Georgia, about the architecture of their houses and the use of the land around them. From Sara, I learned that these architectural styles were brought to the Americas by captured and enslaved people of West Africa. One September afternoon, Sara and I stopped to say hello to Bea Robinson, one of the people whose homes Sara had studied and with whom she had become friends. As they were deep in conversation on the front porch, I was taken by Mrs. Robinson's garden, which was beautiful in the soft, late afternoon light. I asked Mrs. Robinson if I could take some photographs of her yard (Figure 5.3). After a short while, I asked if I could also make a photograph of her, and she allowed me to do that as well.

After our visit, Sara told me that in her research she had learned that not only the architecture of the houses, but the yards' physical layout and use of materials had been traced to West Africa. In the next few days she took me to several other houses and gardens; I fell in love with what I saw.

To my eye and heart, the use of materials—both plant life and material objects— was deeply evocative, although at first, it wasn't easy to articulate why I felt that way. I felt something that was transformative—in the sense that I was in a space that took me out of myself, as happens in the presence of listening to compelling music, entering a silent temple, or sitting on the rocks on the shore in my ancestral home of Prince Edward Island. As I wrote in the preface to my book, *Places for the Spirit: Traditional African American Gardens*:

> these gardens speak a certain language about the earth, about beauty, and about spirit. Some of the vocabulary of this language is about belief and spiritual knowledge—the empty bottles, the pipes sticking upright out of the ground, dolls. Objects have specific meanings that relate to the spirits of ancestors or magical powers, meanings that go back centuries and across an ocean. Some of the vocabulary is functional, practical, born of necessity— the vegetable gardens, the chicken coops—and some is quite simply about beauty—the impatiens and petunias and pinks, the rose bushes, prickly pears, and canna lilies. The way the vocabulary is put together is based on tradi- tion, custom, function, and each gardener's sense of what looks pleasing, in a special and recognizable style. This style becomes the structure of the language; this structure is aesthetic; and this aesthetic, to my eye, is beauty. The language of these gardens has a sound so lyrical that, even though I don't know the nuances of all the words, I could hear it and use it to make these photographs.[2]

To find such gardens, most often alone, but sometimes, especially on longer jour- neys, accompanied by my husband and once by my friend Sara, I drove on country roads and in Black neighborhoods throughout the deep South, looking for gardens that silently called me to stop the car and step out. When I found a garden that held an exquisite coming together of plants and objects, I got over my innate shyness, knocked on the door, and asked if I could photograph their beautiful garden. I ex- plained that I was an artist, photographing traditional African American gardens, exhibiting the images occasionally, and, as time went on, that I hoped to publish this work. I was aware that my whiteness gave me a certain power (whether or not I wanted it, for I could not shed it) and understood that this could make me an un- welcome visitor. But I also thought—or hoped—that as a woman, small and with a quiet voice, that I would not be considered threatening. I hoped that my genuine pleasure and appreciation for the garden was perceived, and though perhaps only tentatively at first, I was nearly always allowed into the garden. When I found a publisher, I wrote to each gardener-owner, sending a few prints and asking for writ- ten permission to publish those photographs.[3] That was followed later with another print and a copy of the book to each gardener.

Once I was invited into the garden, I set up my camera, most often a 4x5 view camera with a tripod and black cloth, but sometimes a medium format camera, and the paraphernalia that comes with using Polaroid materials, which included packs of film, a specially designed bucket for the necessary sodium-sulfite solution to process negatives, and of course a trash bag. My activity was not subtle. I spent an hour or two in a garden—moving around, figuring out points of view, thinking about composition, photographing, and chatting with the gardeners. Typically, the gardener-owner sat on their porch, watched me as I photographed, and we talked as I worked. Using Polaroid materials not only gave me a beautiful fine grain negative, but a positive print of each image I made, and I gave the positives to the gardeners as I worked. In retrospect, I believe seeing my pictures helped to create trust as the pictures clearly showed that I was photographing the beauty of their garden. It was evident that there were no secrets in what I was doing.

In each photograph, I wanted to create a sense of what it felt like to be in the garden, or at least, what it felt like to me. I sometimes also made portraits of the people whose gardens I was photographing. I saw the gardens as artworks, attended to as closely as one might in creating a painting. I attempted to convey this artistry and attention by including what was most visually compelling (the material physical objects, the plants, the organization), framing the image so that everything within it mattered, and by composing to convey my feeling of spatial relationships. Light is a key element in making any photograph—and in these gardens light danced off reflective surfaces, revealed open spaces, and made shadows that created dark contrasting shapes that can intensify the whiteness of an object or lend a sense of mystery and the unknown. Black-and-white photographs present the world abstractly, differently from the way our eyes and minds normally see, as they convert the world of color into the many values of gray. For me, by photographing the gardens in black and white, I created a metaphor for the sense that we are entering a very different place—both emotionally and spiritually.

Far too briefly and incompletely, I will describe what makes the gardens so visibly distinctive. But first, it is important to emphasize that the material objects, the design of the garden, and even the plants themselves represent something other than what they are in themselves. An old tire painted white is not simply an old tire, spruced up to look handsome (though it is that too), it holds several other meanings of a spiritual, ethical, and social nature (Figure 5.6). These meanings are central in a way perfectly described in "The Habit of Art" by Toni Morrison:

> Art is not mere entertainment or decoration … it has meaning, and …we both want and need to fathom that meaning—not fear, dismiss, or construct superficial responses told to us by authorities. It [is] a manifestation of what I believe to be true and verifiable: the impulse to do and revere art is an ancient need.[4]

The meanings of any artwork are subject to interpretation, and the African American gardens are no exception. Gardener-owners who have been interviewed speak of several impulses: they draw on traditions which have both spiritual and personal roots and they speak of the desire to do creative work. Scholars, primarily Grey Gundaker, Judith McWillie, and Robert Farris Thompson, who are white, have traced the philosophical, religious, spiritual, and artistic beliefs and practices in African American yards to those seen especially in the Kongo and Yoruba peoples of West Africa. David W. Jackson, an African American scholar, emphasizes that the gardens preserve memory and identity in a way that includes religious beliefs and personal experiences. The desire to create art in yards as the primary impulse is highlighted by curator Lizetta Lefalle-Collins and writer Alice Walker. My sense is that all of these motives and meanings are manifested in the gardens. Without doubt, the variety of garden designs and constructions shows the intention to be creative, to do artistic work; the preservation of memory and identity, including the recent personal past, is aided and supported by traditions that have lasted centuries, dating back to slavery and before that to Africa.

Underlying what one can see in the gardens are philosophies that came with the forced migration from West Africa of many groups of people—especially Yoruba and Kongo peoples. Although it is beyond the scope of this chapter to explore the complexity and richness of the particular ways in which different beliefs manifested in the American South. The philosophy and religion of the Yoruba persists as a vital force today. In *Flash of the Spirit*, the eminent art historian Robert Farris Thompson describes the art and philosophy of the Black Atlantic; he writes that the highly significant divine force, *àshe*, "bestows upon other gods and spirits, our ancestors," and "upon us the morally neutral power to make things happen, to give and take things away, even to kill and give life."[5] Farris Thompson further explains that in Yoruba thought, *àshe* shows itself in the individual graced with good character, an ensemble of qualities the Yoruba called *itutu*.

> To the degree that we live generously and discreetly, exhibiting grace under pressure, our appearance and our acts gradually assume virtual royal power. As we become noble, fully realizing the spark of creative goodness God endowed us with . . . we find the confidence to cope with all kinds of situations. This is àshe. This is character. This is mystic coolness. All one.[6]

Through *àshe* we are given the power to affect those around us and the responsibility to act with the highest form of integrity. In the world of the Yoruba and Kongo peoples, we exist in deep ongoing relationship with the spirits of our ancestors. With a recognition that both good and evil exist, we must choose how to act. Understanding that we have spiritual guidance and the power to choose how to live is the key to the meanings and symbols of many of the objects, plants, and the overall design of the gardens I photographed.

In traditional African American yards, the garden is meant to be a safe and secure place, therefore to define the space, borders are created and often built with objects such as white-painted rocks, bricks, and tires (Figure 5.4). And yet rarely is there a gate that needs opening, and sometimes there are even welcome mats at the entrance; visitors are clearly welcomed (Figure 5.1). The gardens often include a place in the front yard for social gatherings, another sign of welcome. Still, as we enter, or merely pass by, there are many indicators of how we should behave, indeed warnings that we are being watched. Figures such as birds, chickens, and dolls are placed near entrances and throughout the garden to serve as watchers (Figures 5.1, 5.2, and 5.3). Their presence is meant to remind us to act with discretion and respect when we come into the yard. From the Kongo religion come other traditions seen in the garden; one is the belief that an upside-down or broken vessel, whether a brilliant blue glass bottle, an empty jar, or a clay pot, captures evil spirits to keep them from harming us, as it also contains the spirits of the dead, to keep them from leaving and enable us to learn from them; these vessels are traditionally placed on tree stems and laid horizontally around graves (Figures 5.8 and 5.9). Likewise, a properly swept dirt yard wards off evil as it also creates a sense of calm and peace (Figure 5.2). The color white indicates the divine and represents goodness of character. Pipes of all widths and lengths stuck into the ground allow for communication between us and the spirit world (Figure 5.1). Circles are a sign of progress, signifying forward momentum despite difficulties in life, just as they also refer to the cycles of nature, in which both the living and dead participate (Figures 5.3, 5.4, and 5.6) Robust plants indicate living a responsible life (Figure 5.7). Objects that have been used, especially by family or friends, are spiritually valuable, which something new cannot be (Figure 5.5). Water is both life-sustaining and a threshold between the everyday world and the world of spiritual power, between the world of the living and the dead, and therefore all things related to water are significant—from a water pump to seashells (Figure 5.1). Like water, light and reflectivity are premium signs of the divine; thus anything that creates light such as a mirror, light bulb, or shiny surface has great import. Chairs, stools, and thrones, set aside from sitting areas and sometimes elevated or set off-angle, refer to the spirits of deities or ancestors or a deceased family member, so make a place for the connection between this world and the other world (Figure 5.8). Meaning upon meaning is developed as these objects and symbols are creatively combined in each garden—I think of the power of tall white plumbing pipes lining the welcoming mat-covered path into Louise Daniels' yard or the upside-down white cup inside a glass upside-down jar, with crisscrossed pipes attached, and a chair hung askew in a marvelous construction in Emma Moore's yard (Figures 5.1 and 5.8). Consider the border around Eula Mary Owen's yard made of tires, rocks, and bricks set on an angle—three rows, all painted a pristine white (Figure 5.4). Study the elements in Alexander Bell's garden of carefully raked dirt, multiple watchers, and a white pipe, once a curtain rod (Figure 5.2).

Everywhere around me, as I photographed, were unspoken messages of the importance of how to live and behave in each garden and in this world: how to live honorably, respectfully, appropriately, kindly. These messages are for those who spend time in the garden and, just as importantly, for those who simply pass by. Although it took me many years to learn something of the language of the messages, I somehow felt their power when I entered Bea Robinson's yard in 1987, and each of the other gardens I photographed in the ensuing almost twenty years.

Today, more than thirty-five years after I began the project, I reflect again on whether, like Eudora Welty, I was accompanied by an angel of trust. I would put it in a different way. Knowing more now than I did when I began about the effects of deeply entrenched racist policies, I have a better understanding of the dynamic that may have occurred when doing my work. I am acutely aware that I was likely to have been seen as an intruder into the gardeners' lives, and that my presence could well have provoked anxiety and fear. Yet despite the poisonous legacy of racism, my request to photograph their gardens was met with kindness and willingness. Perhaps it was the fact that I was drawn to their gardens for their beauty that allowed a kernel of trust to take root. Whatever it was, I am grateful to each of the gardener-owners for taking a chance with me. In turn, I understood that my role as a visitor, as a human, was to be respectful, to honor each gardener, and to see and listen and value everything in the garden. My role as an artist was to create beauty about something that itself was beautiful and that holds such significant cultural and historical meaning.

I have come to believe that in a very real sense, it was the gardens themselves that provided the foundation. The messages from the spirits of the ancestors, the reminders of living properly and respectfully, the presence of upside-down bottles and cups keeping evil away and circles signifying that progress will occur—all this and more truly did create a safe and secure space. Trust came because of those spirits. Each gardener and I, their visitor if only for a few hours, knew we were safe, together.

Acknowledgments

My deepest gratitude goes to all the gardeners for creating the gardens and for allowing me to create photographs of their gardens. My great appreciation also goes to a number of other people who helped me in this project, especially my husband Lowry Pei and close friend Sara Glickman, but I can only name a few here. From a number of scholars and writers, both African American and white, I learned the multiple ways one can understand and appreciate the historical and artistic significance of the traditional African American garden. From Richard Westmacott and Sara Glickman's work, I learned a good deal about the history of African American yards in the rural South. I learned about the West African spiritual and philosophical beliefs evidenced in the objects and design from three additional scholars: Robert Farris Thompson, Grey Gundaker, and Judith McWillie. David W. Jackson III's thesis, a case study of one remarkable African American garden, led me to other theories of how to understand the gardens, especially the work of curator Lizetta Lefalle-Collins. My thanks also to several readers who have been helpful with this chapter, Carl Glickman, Katherine Jelly, Lowry Pei, Pam Swing, and my cohort at Brandeis Women's Studies Research Center (WSRC). Searching for a meaningful way to talk about my personal experience working on this project, I found what felt like the right model, developed by K. Melchor Quick-Hall, an African American scholar and my colleague at Brandeis WSRC. She proposes a principle of "radically transparent author positionality" to link scholarship (or in my case artwork) to the author. My continued appreciation goes to my editor Barbara Ras who believed so deeply in the value of this work and to Trinity University Press for publishing *Places for the Spirit: Traditional African American Gardens*.

I am also grateful to the viewers of my photographs who continued to teach me about the gardens. Because I was most intent upon photographing, I did not ask the gardener-owners about the traditions of objects or design, although occasionally they spoke of carrying on in the tradition of a parent or told me a story of their family that dated back to slavery, so I do not know what specific objects, designs, and plants meant to each gardener. However, once I began exhibiting and giving talks, African American viewers often confirmed what I had read and told me things I hadn't known—such as the pattern in which the is earth is raked is significant as it can ward off evil and the color blue has protective powers as it frightens away ghosts. One woman told me that when she was a child her parents would go inside and pull down the window shades when they intended to talk about the spiritual aspects of their garden, so that white people would not see and hear them. She also told me, as did many other African American viewers, that she was grateful for my doing this work. I felt she was glad the window shade had been raised. And finally, another African American colleague at Brandeis WSRC, Edith Coleman Chears, said to me recently, whether I had the right to do this work is not the point—rather, having done it, I had the responsibility to share it.

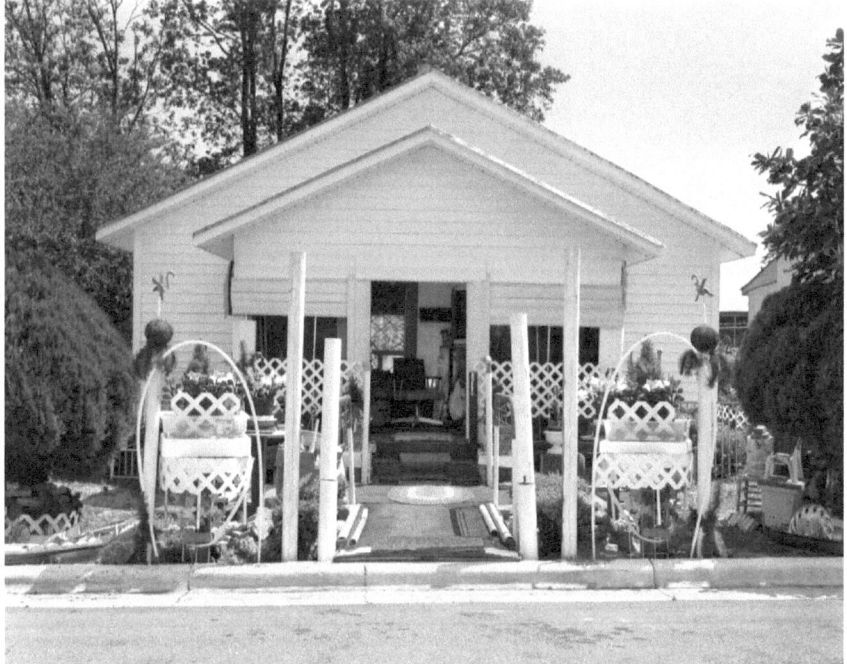

Figure 5.1 Louise Daniels' Garden and House, Greenville, North Carolina, 2005.
An elaborate symmetrical entrance to both the yard and, with welcome mats invitingly placed, the sparkly white house was created by a variety of pipes— standing, curved, horizontal—that framed the path to the door. Constructions in the yard use myriad objects, including reflective hubcaps, two flamingos, bottles placed upside down in the garden, and white lattice work. All this and more was set among carefully trimmed shrubs and potted plants.

Source: ©Vaughn Sills, from *Places for the Spirit, Traditional African American Gardens*, Trinity University Press.

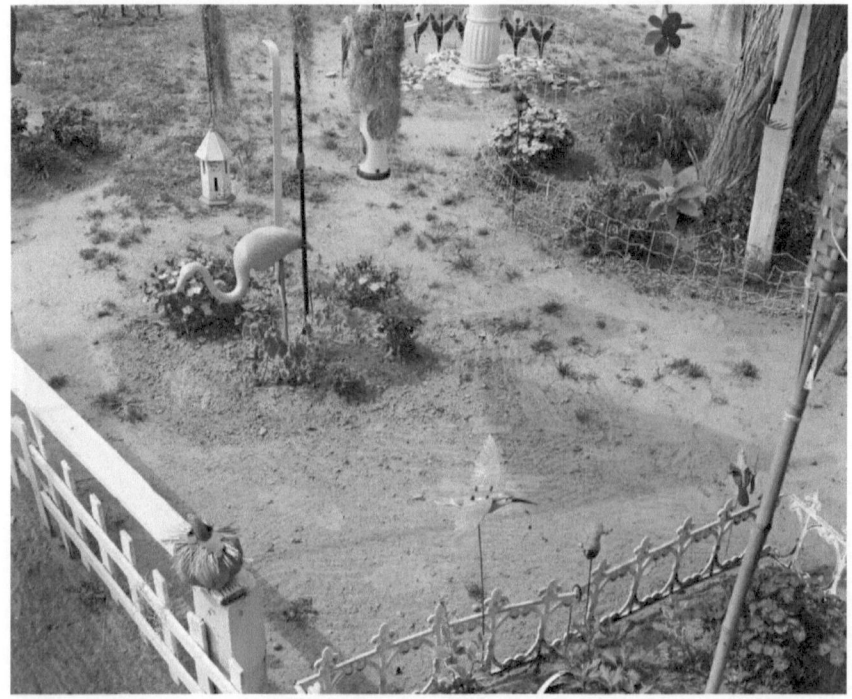

Figure 5.2 Alexander Bell's Garden, Greenville, North Carolina, 2005.
Every square inch of the small yard in front of Alexander Bell's home was gardened with attention to detail. Figures of birds, a curtain rod, whirligigs and small plants, both real and artificial, were balanced by the clear open space of raked earth.

Source: ©Vaughn Sills, from *Places for the Spirit: Traditional African American Gardens*, Trinity University Press.

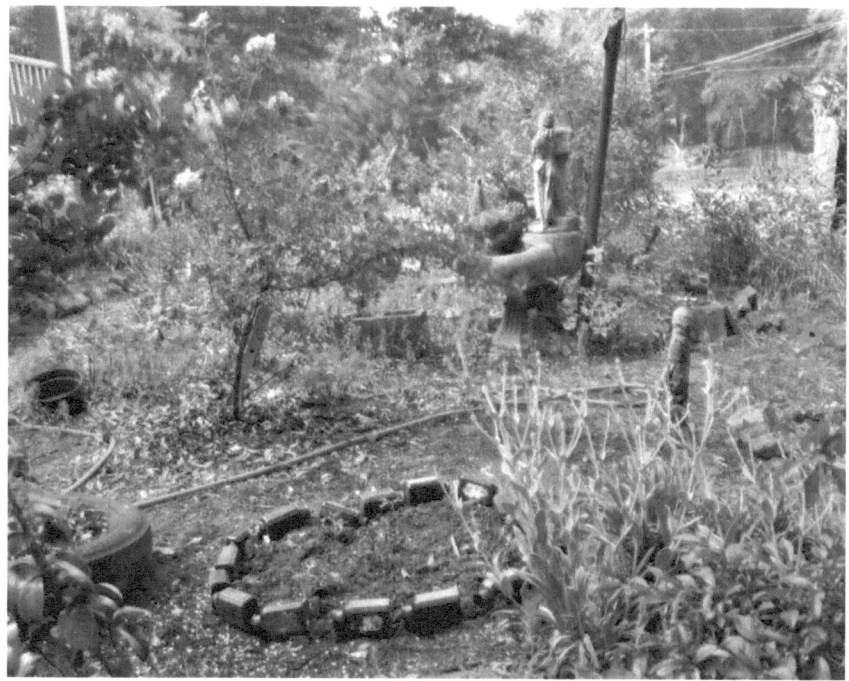

Figure 5.3 Bea Robinson's Garden, Athens, Georgia, 1987.
Light glowed through Bea Robinson's garden; it reflected off brown bottles laid in a circle and bathed two sculptures, a vaguely mythological creature and a chicken. Nearly the same size and evidently friends, they seemed to watch passersby and visitors with equal dignity. Near them a fat pipe, wrapped in cloth, emerged from the ground.

Source: ©Vaughn Sills, from *Places for the Spirit: Traditional African American Gardens*, Trinity University Press.

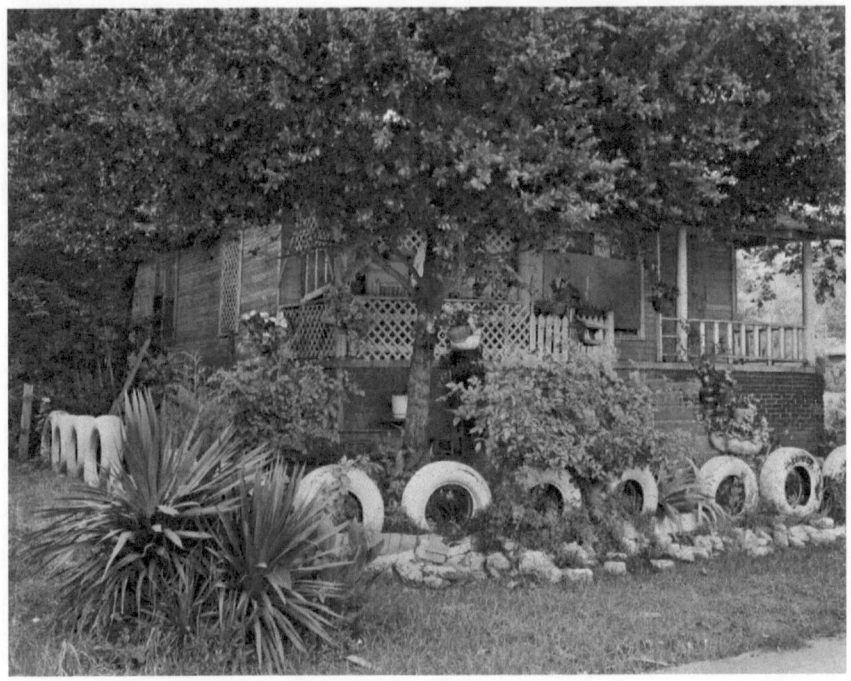

Figure 5.4 Eula Mary Owen's Yard, Jackson, Mississippi, 2002.
Under the wide spread of a cherry laurel Eula Mary Owen tended a shady garden inside an impressive border of white painted tires, rocks, and bricks. Inside the garden were white painted tires, planters, and many pots holding both natural and artificial flowers, as well as plastic chickens and a bird feeder.

Source: ©Vaughn Sills, from *Places for the Spirit: Traditional African American Gardens*, Trinity University Press.

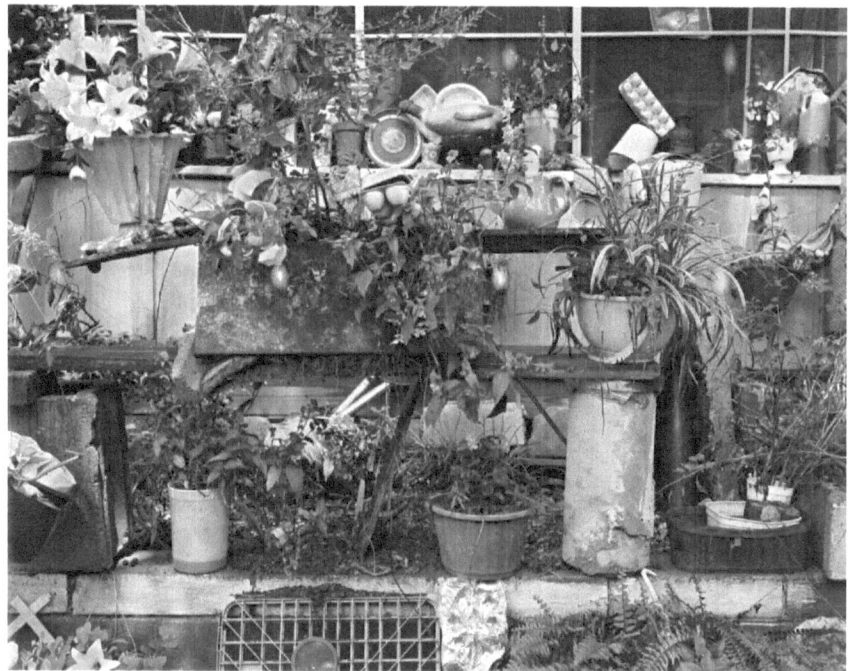

Figure 5.5 Annie Belle Sturghill's Garden, Athens, Georgia, 1988.

Annie Belle Sturghill's garden filled the length and depth of the land in front of her house. She worked on her sculptural assemblage as one would a painting, always considering the overall composition as well as the parts. The elements are too numerous to name but include cloth, eggs, jars, and bottles, Styrofoam peanuts, wheels, the reflective surface of a foil baking pan, and figures of birds— all placed among flowers and foliage. Ms. Sturghill saw her garden as a work in progress, and each day she would add, take away, and rearrange objects in it.

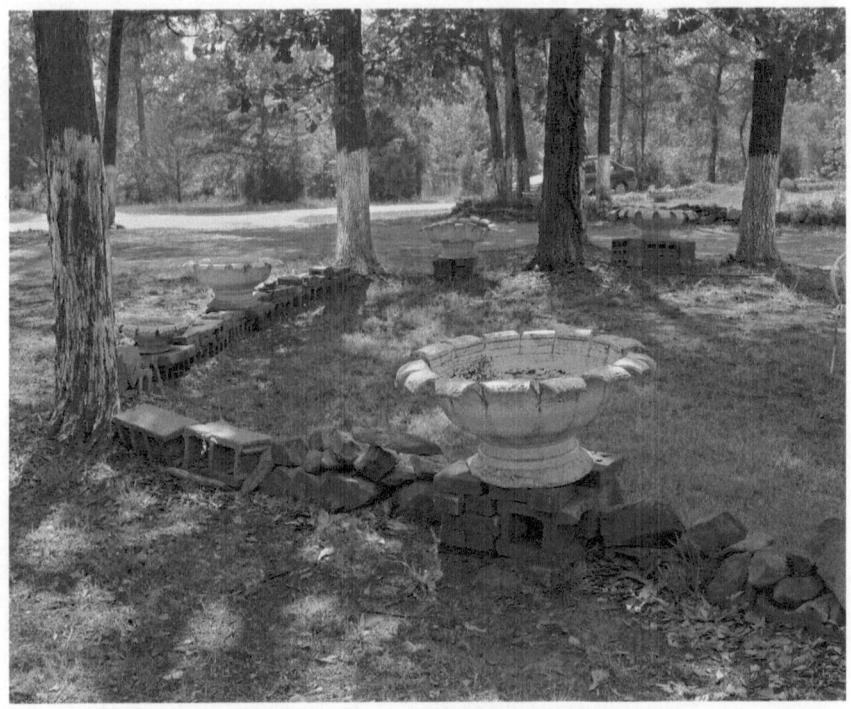

Figure 5.6 Herman Thompson's Yard, Jenkinsville, South Carolina, 2005.
    In his front yard on a serpentine border of bricks and concrete blocks, Herman Thompson placed six urns, sculpted from tires and painted white, amid towering trees painted white at the base. The effect was ethereal.

Source: ©Vaughn Sills, from *Places for the Spirit: Traditional African American Gardens*, Trinity University Press.

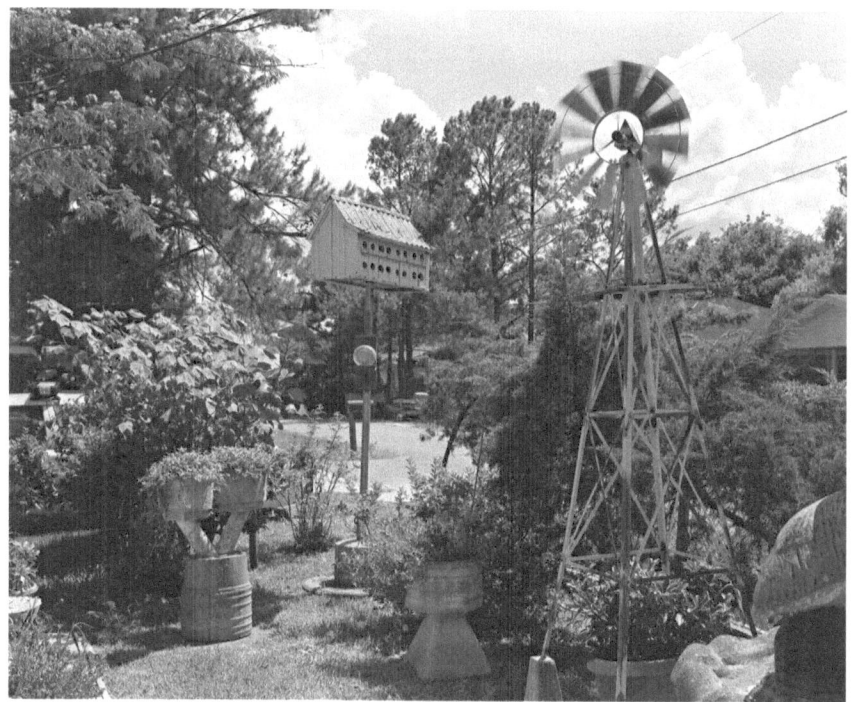

Figure 5.7 Jules Landry's Garden, St James, Louisiana, 2005.
     Along a quarter-mile residential street off River Road which parallels the Mississippi River, nearly every garden held monumental cement planters unlike anything I'd seen before. The Rev. Jules Landry designed and built these three-foot objects holding one, two, or three plants, using forms he got from the nearby factory where he worked prior to his retirement. The solidity of these planters contrasted delightfully with an airy windmill and a birdhouse.

Source: ©Vaughn Sills, from *Places for the Spirit: Traditional African American Gardens*, Trinity University Press.

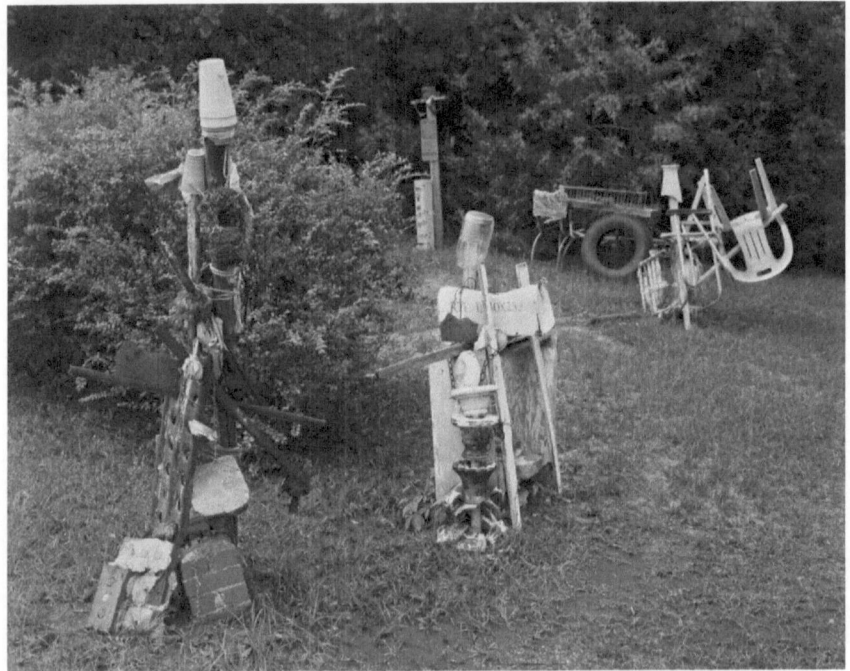

Figure 5.8 Emma Moore's Yard, near Marion, Alabama, 2001.
　　　Three sculptural constructions drew my attention to Emma Moore's yard.
They were made of cups and bottles, pipes, circular objects, a plastic chair, and
many other objects tied together in unexpected combinations.

Source: ©Vaughn Sills, from *Places for the Spirit: Traditional African American Gardens*, Trinity University Press

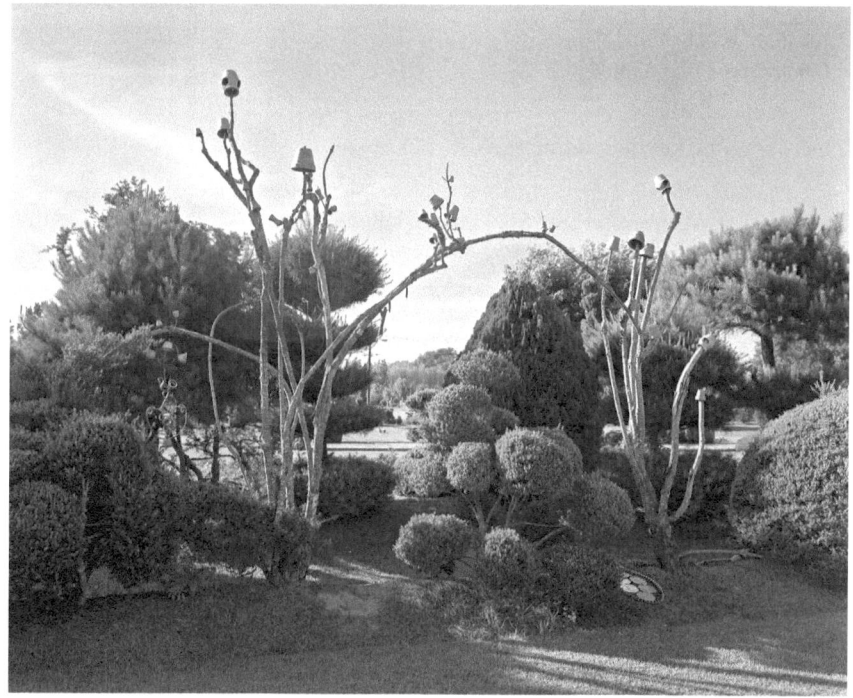

Figure 5.9 Pearl Fryar's Garden, Bishopville, South Carolina, 2002.
    Although Pearl Fryar's extraordinary yard of topiary at first seemed unchar-
acteristic of traditional African American gardens, the construction of inverted
pots on the ends of bare branches felt similar to a bottle tree, and it artfully
contrasted with the topiary's density and volume. Mr. Fryar told me he was influenced by his mother's and his grandmother's gardens and is working in their
tradition.

Source: ©Vaughn Sills, from *Places for the Spirit: Traditional African American Gardens*, Trinity University Press.

Notes

1 Welty, *One Time One Place*, Foreword.
2 Sills, *Places for the Spirit,* Preface.
3 In this chapter, I use the term gardener-owner and gardener interchangeably, as the gardeners were, it appeared, owners of their property, they had lived on this land and in their homes long enough to create and establish gardens and would likely be able to stay there. Home-ownership, not always allowed for several reasons, is significant in African American history.
4 Morrison, "The Habit of Art," 54.
5 Thompson, *Flash of the Spirit,* 5.
6 Thompson, Flash of the Spirit, 16.

References and Further Reading

Arnett, Paul and William Arnett, eds. *Souls Grown Deep, African American Vernacular Art*. Vols. I and II. Atlanta, GA: Tinwood Books, 2001.

Glave, Dianne D. *Rooted in the Earth: Reclaiming the African American Environmental Heritage*. Chicago: Chicago Review Press, 2010.

Glickman, Sara. *Historic Resources in African-American Neighborhoods in Piedmont Georgia*. Athens: Master's Thesis, University of Georgia, 1986.

Gundaker, Grey. "Tradition and Innovation in African-American Yards," *African Art*, 26 no. 2 (April, 1993): 58–71, 94–96.

Gundaker, Grey, ed. *Keep Your Head to the Sky, Interpreting African American Home Ground*. Charlottesville: University of Virginia Press, 1998.

Gundaker, Grey and Judith McWillie. *No Space Hidden, The Spirit of African American Yard Work*. Knoxville: University of Tennessee Press, 2005.

Hall, K. Melchor Quick. *Naming a Transnationalist Black Feminist Framework: Writing in Darkness*. London and New York: Routledge, 2020.

Jackson, David W. III. *The Walking Nkisi: African-American Material Culture in Iowa, a Case Study in Yard Art in Waterloo, Iowa*. Iowa City: Doctoral Thesis, University of Iowa. 2006.

Lafalle-Collins, Lizetta: *Home and Yard: Black Folk Expressions in Los Angeles*. Los Angeles, California Afro-American Museum. 1987.

Morrison, Toni. "The Habit of Art," in *The Source of Self Regard*, New York: Vintage Books, 2019: 54–57.

Sills, Vaughn. *Places for the Spirit: Traditional African American Gardens*. San Antonio: Trinity University Press, 2010.

Thompson, Robert Farris. *Flash of the Spirit, African and Afro-American Art and Philosophy*. New York: Vintage Books, 1983.

Walker, Alice: "In Search of Our Mothers' Gardens," in *In Search of Our Mothers' Gardens*. New York: A Harvest Book, Harcourt, Inc., 1983: 231–43.

Welty, Eudora. *One Time, One Place*. New York: Random House, 1971.

Westmacott, Richard. *African American Gardens and Yards in the Rural South*. Knoxville, University of Tennessee Press, 1992.

6 On the Diagonal, through the Window

Marie Menken's *Glimpse of the Garden*, 1957 and Rosalind Nashashibi's *Vivian's Garden*, 2017

Maureen Turim

If plants can be a living palette, if gardens can be art, carved in a landscape, and if gardens juxtapose shapes and colors to create compositions (sometimes balanced, highly arranged, and orderly, sometimes angled and edgy), shooting an artistic film in a garden is a more daring undertaking than might first be evident. The challenges are greater than those of the still photographer, whose work in the garden dates from early photographs,[1] and whose historical artistry in this domain has recently been explored in Jamie M. Allen and Sarah Anne McNear's *The Photographer in the Garden* (2018). In comparison, the filmmaker, working with time as well as space, needs to do more than reiterate the space itself—its pathways and composed vistas. Unless film limits itself to pure documentation, it necessarily transforms the beauty of the site into a different aesthetic undertaking as it reframes and rearranges space within a temporal unfolding. Marie Menken's 16 mm, five minute, color film, *Glimpse of the Garden*, is a foundational work of US experimental filmmaking from 1957 that undertakes just such a creative reframing in ways that were deeply innovative at the time. Menken's painterly eye frames abstractly the spaces of a large garden extending into grassy knolls. The first part of this chapter will explore her film's exploration of visual pleasure and dynamic aesthetics. This work is often seen as a prime example of personal filmmaking, by which is meant both that the individual filmmaker conceived, shot, edited, and scored the film herself, and that the film expressed her interior state, her vision, her singular perspective. While this is certainly true of Menken's film in most regards, I want to also look at the film as a collaborative effort, in this case with her friend whose garden is filmed.

More recently, Rosalind Nashashibi's experimentally conceived documentary, *Vivian's Garden,* from 2017, narrates the lives of a pair of artists, Elisabeth Wild and Vivian Suter, in a Guatemalan garden enclave, as a refuge from historical political displacement. Wild is Suter's mother, and the close relationship of these two creative women provides the impetus for the weaving of their lived experience with the visual textures of the garden and the artworks. Their artwork is shaped by the way they have collaborated in taking care of each other throughout their lives, being each other's first audience for the works they create. Nashashibi constructs a film that speaks to these artists' work by achieving its own aesthetic life. Thus, this

DOI: 10.4324/9781003381549-7

film may be considered collaborative art on numerous levels: between the two artists, between these artists and the filmmaker, and between all three and the garden
surround.

The two films, made sixty years apart, represent two different paradigms of
women's garden-inspired artmaking. Discussing them together will allow me to
propose a feminist relocation of the garden setting as the exploration of interior-
ity, lived experience, and aesthetic force. While the two films belong to different
genres of filmmaking, experimental or avant-garde cinema in the case of Menken's
and documentary filmmaking in the case of Nashashibi, these two genres have a
long history of interspersed techniques and goals. Today we often speak of "experimental documentaries" and of the documentary component to such works as Stan
Brakhages's *The Wonder Ring* (1955), shot from New York City's Third Avenue El
just before that elevated line that had become part of the IRT (Interborough Rapid
Transit Company) was demolished. In addition, Soviet montage, an experimental
technique that finds it echoes in the works of both of the women filmmakers I address here was developed across genres in the films of Sergei Eisenstein, Dziga
Vertov, Ester Shub, Viktor Turin, Alexander Dovzhenko, and Vsevolod Pudovkin.
Considering Menken's and Nashashibi's films side-by-side allows us to see how
Menken's work effectively documents the garden of her friends Dwight Ripley
and Rupert Barneby, while Nashashibi's work uses evocative montage of sound
and images to signal its affinity with the art works and the garden space, ascend-
ing to the goals of visual and auditory pleasure and expansiveness it shares with
experimental works.

"Avant-gardens" is Scott MacDonald's provocative punning title for a chapter that links the filmic avant-garde to the garden in his book, *The Garden in the
Machine: A Field Guide to Independent Films.* In this chapter he discusses Marie
Menken's *Glimpse of the Garden* alongside other avant-garde film works shot in
gardens, or in the case of Rose Lowder's *Les Tournesols*, 1982, in a commercial
sunflower field. In MacDonald's opinion Menken's film is about "place," as he
emphasizes how this experimental filmmaker interacted with the site of her film to
produce a work that would influence other filmmakers, notably Stan Brakhage and
Jonas Mekas. He situates the film in a biographical frame, which includes Menken's
marriage to fellow filmmaker Willard Maas, and her role in the New York art scene
of the fifties and sixties. Admittedly, the biographical elements are fascinating in
their own right, as Maas and Menken negotiated their marriage in the context of
Maas's bisexuality, as well as his founding of an art movement, Gryphon. Recently,
Juan A. Suarez, in "Myth, Matter, Queerness: The Cinema of Willard Maas, Marie
Menken, and the Gryphon Group, 1943–1969," provides a scholarly analysis of
this movement in the context of a contemporary queer perspective, making the bio-
graphical elements relevant to the understanding of the artmaking. However, Mac-
Donald relies on Stan Brakhage's chapter, "Marie Menken," in *Film at Wit's End*;
Brakhage remarks that the garden in which Menken filmed was created by Dwight
Ripley, a man whom Brakhage wrongly claims was a former lover of Menken's
husband (he wasn't, though his partner may have briefly had an affair with Maas).

Brakhage's approach hovers voyeuristically over lives that today might be better understood through the consciousness of feminism and queer theory.[2]

Thus, my methodology for looking at the film differs. First, I will avoid this personal biographical approach, partially because Brakhage, despite having been a historical witness who knew Menken, gets some of the biography wrong. Brakhage's tendency to opt for sensational stories over a more measured assessment will come as no surprise to those familiar with his work, *Film Biographies*, which takes great license with the lives of such figures as Sergei Eisenstein and Friedrich Wilhelm Murnau. MacDonald relies on Brakhage's account, with consequences I will examine shortly. Second, I will argue that early writing on the film by male authors neglects a feminist context for understanding women's role in the avant-garde. Menken's work may be seen as linking gender, gardens, and pleasure, infusing her short film with contrasts of shape and color. These striking visual statements recall her previous work as a painter and are deeply connected to her other films that transform objects into light abstractions. Key to this film, like her others, is the juxtaposition of filmic shots in motion and montage. Throughout the film there is a tension between the objects filmed and the abstract gestures of the camera. Those gestures steal the objects away from any exclusive sense of being fixed objects. Rather, the flower and plant objects are seized as elements of visual, temporal, and rhythmic compositions for which camerawork and editing provide the strokes. The film glimpses the garden as an experience subject to such transformations, rather than a place fixed in space.

In checking for the authenticity of the biographical material that Brakhage and MacDonald offer, I found some salient background material—primarily in the archives of the New York Botanical Garden—that is particularly relevant to consideration of the garden setting.[3] MacDonald, following Brakhage, misidentifies the location of the garden and leaves out interesting background on the garden's provenance that is particularly relevant, as this garden is steeped in botanical science. The garden indeed belonged to Dwight Ripley, as Menken's hand-designed credits indicate, but their accounts situate the garden as located at a subsequent Ripley home on Long Island. In fact, the filming location was an estate in Wappinger Falls, Dutchess County, New York, where Ripley lived with his partner, British émigré Rupert Barneby. There they built a rock garden of considerable note, though it is Ripley who designed and tended the garden. Both men were already established horticulturists in Sussex, England, before moving to Los Angeles in 1939, and then to New York State in 1943. Barneby and Ripley had collected plants in North America as early as 1936 and were responsible for identifying for the first time many plants of the western United States and in Mexico, some of which would vanish from their natural habitat due to A-bomb testing at the Trinity bomb site in central New Mexico in 1945. Dwight Ripley was also an artist of note, whose work had five solo exhibitions at the Tibor de Nagy Gallery beginning in 1951. Six of his drawings were included in an exhibition at Peggy Guggenheim's legendary gallery, Art of This Century. Marie Menken made a film, *Dwightiana* (1957), that playfully explored his paintings. In an excellent essay, "Swing and Sway: Marie Menken's

Filmic Events," Melissa Ragona analyzes this film that is evocative of the same imaginative artistry we see in *Glimpse of a Garden*:

> Menken uses animation to "'move'" the image in unexpected, novel ways. She begins this piece with paint dripping down over blue and black title designs— these drips will appear quite literally, again, in *Drips in Strips* (1963). But here paint's heavy gravity is juxtaposed against animation's ephemeral agility. First, Menken syncs each drip with a percussive stroke on a talking drum from accompanist Teiji Ito. This opening tableau is followed by the animation of a kaleidoscope of brightly colored objects moving over one of Dwight Ripley's Miro-like paintings which exude a kind of magic realist aesthetic (griffin-like figures move in a surreal garden). Here, as she does in her other painting-related films, Menken comments on the use of foreground and background, screen and frame, as well as 3-D versus 2-D space. Ripley's paintings work both as flat planes, exposed as painterly surfaces, and as open fields in which animated objects enter or scurry across in agitated, jazzlike patterns. Menken uses sand animation to further decenter the picture plane of each painting, rearranging focal points through a system of "'cover up'" and "'reveal.'" Then, objects—necklace strands, bits of jewelry—take command and seem to be consuming their sand background as they move across the screen. Studies of stasis versus movement, aggregate versus solo constellations dominate the film, accentuated by Teiji's insistent music.[4]

Rupert Barneby would later work as a researcher at the New York Botanical Garden, which holds his papers and those of Ripley. A number of species were named after Ripley: *Cymopterus ripleyi, Aliciella ripleyi, Astragalus ripleyi, Eriogonum ripleyi, Omphalodes ripleyana* and *Senna ripleyi*, the first three of which he co-discovered with Rupert Barneby. Barneby has been acclaimed one of the world's leading plant taxonomists. His main areas of expertise were the *Leguminosae* and *Menispermaceae*. He described and named 1,160 plant species new to science. Five new genera and twenty-five species were named in his honor. Barnaby and Ripley's life together has recently been chronicled in *Both: A Portrait in Two Parts* by Douglas Crase. As partners in both life and botany, then, Barnaby and Ripley provide not just any garden to their friend Marie Menken, but a living laboratory layered with scientific and preservationist vocations, a curious collection of the rare and unusual, where the delicate blooms are distinctly different from the types of flowers and shrubs found in more classical gardens. Their low-lying succulents, which Ripley cared for in a large greenhouse, provide their own carefully potted arrangements to be explored in panning shots and close-ups. One of the innovative aspects of Menken's film work here is use of a magnifying lens in front of her Bolex's close-up lens. Her film is truly experimental in that it apparently approximated the macro lens work by artisanal means.

On a more theoretical note, consider that MacDonald's chapter appears in his book whose main title, *The Garden in the Machine* evocatively inverts the main title of Leo Marx's famous work *The Machine in the Garden*. This 1964 study engages with the pastoral fantasy that colors US ideology and plays a role in shaping US literature. The garden that figures in Leo Marx is metaphorically the Garden of Eden: it is wild nature. Technology's entrance into this picture is described by US writers, bearing ever more consequence. Marx writes the "recurrent image of the machine's sudden entrance into the landscape" was characteristic of American writing by the turn of the twentieth century.[5] As agriculture and gardening are technologies that harness and transform nature, the garden that meets the mid-century machine, the portable camera, in Menken's film is a studied construction, just as the camera is a device that mediates human vision. In Menken's hand, the camera intrudes and wanders, with a caressing gesture that treats the plants as objects that shine: they shape, color, and define the filmic frame. The question for Menken as artist is not simply how can my camera show this space, but what can my camera do to delineate a series of transformative visual framings inspired by the objects before it? The spatiotemporal series of frames interacts with the shapes of these plants against their grounds, in succession as the camera moves. We never forget that the glimpse we see in this film is a camera-mediated look that revels in its own creative power. Later in the film some of the plants earn individual solos and extreme close-ups.

This work of the camera shall be paired with cinematic editing: in the process, the actual place becomes reconfigured at least in part as fantasy. Menken's film is as much about the camera's ability to abstract and redesign as it is about witnessing and documenting the reality of vision. Menken's energetic pursuit of abstraction is what MacDonald misses when he suggests at one point that the sharp intrusions of camera movement may be seen as mimicking bird flight through the garden; though birdsong punctuates the soundtrack, it is sampled from a stock source. Though zoomorphic vision may be among a viewer's associations, in a larger sense movement here resists any natural, organic, or transparent motivation or reference.

Rather, my take on the film's camera movement is aligned with Ragona's essay on Menken's other films, "Swing and Sway: Marie Menken's Filmic Events." The works discussed are films that stem from Menken's interaction with the art world: Noguchi, Andy Warhol, and Kenneth Anger. Due to this focus, Ragona barely addresses *Glimpse of the Garden*, yet the phrase "swing and sway" which she credits to Parker Tyler, indicates that Ragona's way of characterizing Menken's handheld camerawork can serve our analysis of Menken's garden film, if we take swinging and swaying as far more controlled here than in some of the other films. Camera movement in *Glimpse* includes often angular pans that slice at geometric configurations within the duration of a shot. Temporality is orchestrated in relationship to these angles. Panning shots become vectors, or curvilinear flourishes, and once, as I discuss below, an unusual L shape. Each camera movement has a sculptural component as it carves out space in relationship to that which is figured within the sequential frames.

The opening of the film features the rock garden. In some shots the rock garden seems to glow, especially in the closer shots, as the terrain reflects the sunlight. It is here that we find the astounding camera movement, which elsewhere in analyzing Hollis Frampton's film, *Summer Solstice*, from 1968, I have called an L-shaped

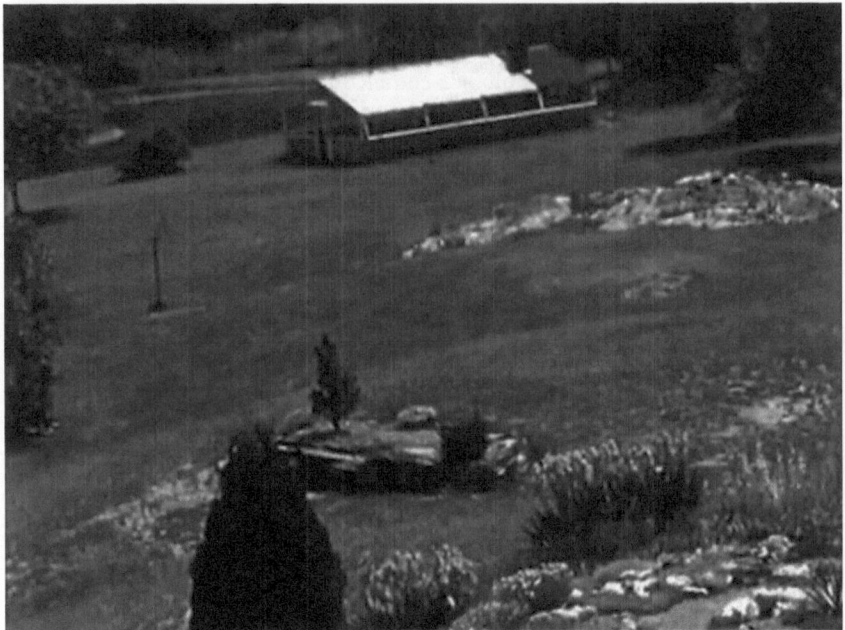

Figure 6.1 The long shot at the end of a tilt-up camera movement, with the greenhouse in the distance. Marie Menken, *Glimpse of the Garden,* 1957, frame enlargement.

shot for the way it carves out space.[6] Here the L is sideways, beginning with a horizontal arcing pan that then abruptly tilts up. Once it has achieved its tilt, it features the only architecture to be seen in the film, the extensive greenhouse in the distance.

Notable in this shot is the great range of the lens Menken is using as her camera shifts in one shot from a focus on the rocks and flowers, to extreme distance, but without the use of a variable lens. Considered less technically, the shot links spaces, grounding us concretely in a longshot for only the briefest of glimpses. No sooner grounded, a cut has the film fly off in a rapid abstract movement to traverse at a diagonal a space dense with vegetation.

Similar diagonals are used in motion to pan across rows of pots containing succulents and other small plants in the greenhouse that we see in the distance in the longshot that ends the tilt up I discussed above.

First, one diagonal moves to the upper right, which is followed by a flowing diagonal in the opposite direction. A tension builds between the rows of round planters and the diagonally sculpted framings. Circles within square frames are fairly rare in cinema, and always notable to me when they occur: consider an image from Sergei Eisenstein's *Strike,* 1925, in which the Lumpen Proletariat emerges from circular holes in the ground to be hired to break the strike. Menken's potted plants may be less foundationally extraordinary, but her rendering of them is just as replete with *ostragene*, the Russian Formalist term for "making strange," renewing a figure by the manner in which it is presented.

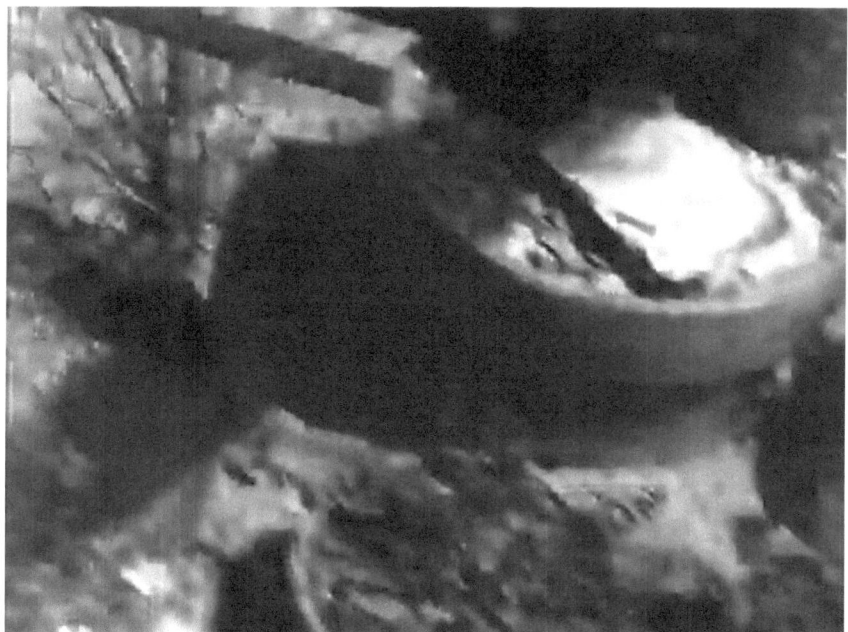

Figure 6.2 Close-up on the potted plants from the diagonal camera movement. Marie Menken, *Glimpse of the Garden*, 1957, frame enlargement.

Another visual flourish is a series of three jump cuts forward on the axis of action on an orange flowering plant. Then a tilt up a spiky green plant chronicles the plant's protrusions that catch the light. A montage of tight close-ups on colorful flowers is juxtaposed with those on plants whose appearance recalls sea creatures and sea vegetation. *Glimpse of the Garden* has moments of darkness, and strangeness, at times sexually suggestive, while always providing a sensuous display. An extreme close-up on a yellow flower recalls the attention devoted to flowers' enticing shapes in paintings by Georgia O'Keeffe, such as *Jimson Weed* (1932) and *Rose* (1957).

In contrast to Menken's abstracted garden, *Vivian's Garden* offers a contemporary filmmaker's view of a garden inhabited by artists; the 16mm film (though often shown in digital transfer) is sensitively framed to explore the dappled light and deep shadows of tropical vegetation, though its imagery is more contextually situated than Menken's rhythmic fantasia. Nashashibi offers us everyday scenes in the lives of two émigré artists Elisabeth Wild and Vivian Suter in their adjoining houses built in a garden enclave in Panajachel, a remote town in the rainforest outside Guatemala City. Elisabeth Wild's life began in Vienna in 1922. Her father, a Jewish wine merchant, took the family from there to Argentina to escape Nazism. As a young woman, Elisabeth worked in textile design, eventually marrying August Wild, the owner of a textile concern. But once again politics, in this case the coming to power of Juan Perón's government, necessitated expatriation, so the Wild family sold their holdings and moved back to Europe, to Basel in 1962.

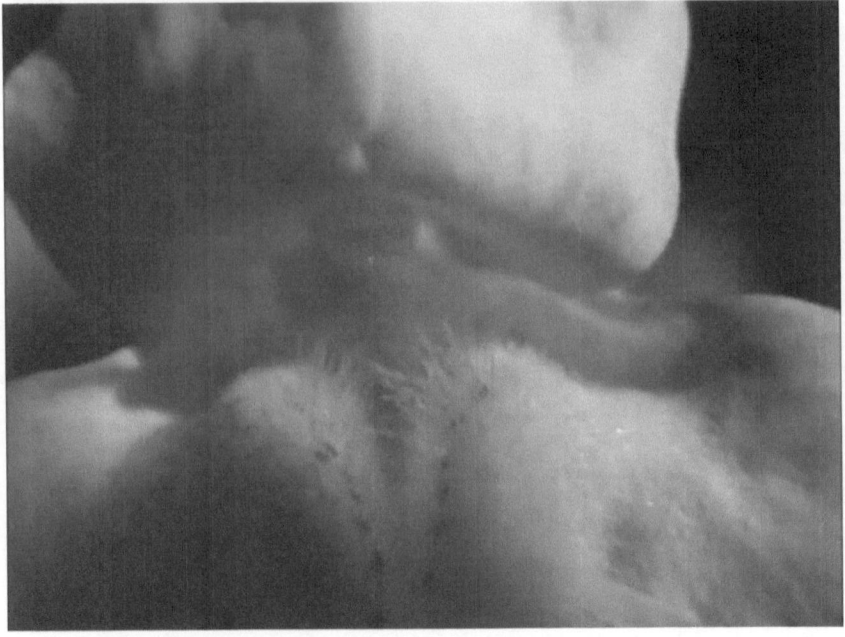

Figure 6.3 Extreme close-up on a flower. Marie Menken, *Glimpse of the Garden,* 1957, frame enlargement.

In 1996, Elisabeth returned to Latin America with her daughter Vivian to live in Panajachel.

Wild's recent practice is collage, made from cutting out and juxtaposing shapes from magazines, and we see her at work at the very beginning of the film, as well as in a brief reprise later. Her daughter Vivian was inspired through painting with her mother to become an artist. She studied in Basel when they lived in that city, and that is where she first began showing her work. She makes color abstractions in oil and acrylic. As Stefan Benchoam, co-founder of the gallery Proyectos Ultravioleta in Guatemala City has explained, Vivian's art practice shifted after Hurricane Stan in 2005.[7] During the mudslides caused by that devastating hurricane, Vivian rescued her paintings from the inundation, then grew to appreciate the traces raging weather had left on her works. Since that time, Vivian has purposely left her paintings exposed to the garden to get an imprint or finishing patina from the elements and the shedding foliage that she allows to interact with them.

She refuses to use elegant frames when she shows her work in galleries and museums, as the weathered paintings are meant to be seen raw, whether on large, thick paper or on canvas. Vivian collaborates with nature, and foregrounds this in the manner she exhibits her works.

The film about Elisabeth and Vivian was commissioned by Documenta 14, the Kassel, Germany-based art exhibition that takes place every five years, each

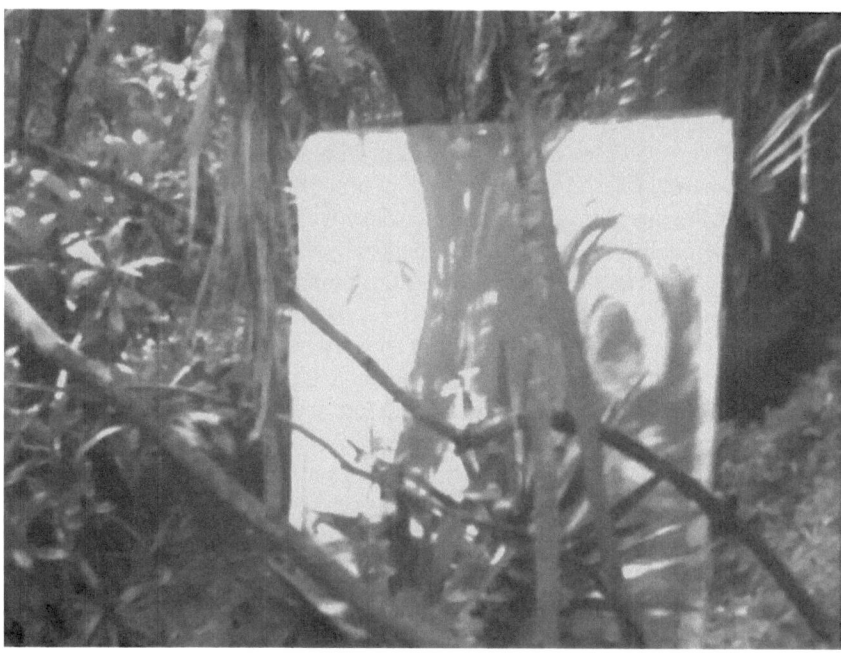

Figure 6.4 Vivian Suter's painting placed in the garden, collaboration with the elements. Rosalind Nashashibi, *Vivian's Garden*, 2017, frame enlargement.

iteration receiving a number; Documenta 14, exceptionally, also had a venue in Athens, Greece, prior to Kassel in 2017. Adam Szymczyk, the artistic director of this iteration became interested in Vivian's paintings from his years serving as director at Kunsthalle Basel, where he discovered a catalogue that featured her work as a young artist in Basel.[8] *Vivian's Garden* originated as a project to contextualize the paintings for the international art world, but with artistic freedom granted its filmmaker to make a work of art in the spirit of the artists whose artmaking process she depicted. Documentaries about artists, especially those commissioned to coincide with exhibitions, are their own subgenre of films; their mission is to supply background biography, studio visits, and art historical evaluations for visitors to the museum or gallery to supplement wall placards, photographs, printed handouts, or catalogues. Here redundancy with the direct explanations those other annotations offer is largely eschewed. Instead, the film observes the artists in their home environment, telling their stories in indirect form and with fragmentation. It is meant to whet our appetite to know more; this film is meant to work as art in parallel to that art alongside which it was first shown. In fact, in the Kassel exhibition space, the film purposely was shown in its own viewing space, at a distance fron the artists' works. At the Chicago Art Institute, on the other hand, the spaces of projection and exhibition while separate, were more closely linked to take advantage of the architecture and philosophies of linking film, photography, and the plastic arts characteristic of the Contemporary Wing of that museum.

In an interview with film theorist and filmmaker Laura Mulvey, Nashashibi explains that the making of this film resulted from three visits to the artists' Guatemala home, the first of which was without cameras.[9] This approach to getting to know her subjects in their environment well before intruding with her camera shows in the finished film, as there is an ease with her presence and the project of filming. Clearly, Nashashibi makes a work of art of the film itself, exploring in poetic imagery the interplay of the garden enclave with the creative processes of the two women. Equally, their artmaking is marked by their deep bond with each other, and the film strives to allow us to witness this human sympathy, while also framing their lives with geopolitical astuteness. Featured are the Guatemalan men and women who work for them, and this from the very beginning of the film. I see this inclusion of working-class Guatemalan laborers to be central to the film's functioning with a political significance that I will explore shortly, but first I want to discuss specifically how the appearance of these laborers occurs in the film.

The film opens with entrance of caretaker, Don Tomás, through the iron gate into the forested garden. It follows him walking the earthen path first to Elisabeth's house, then back outside to traverse the path to Vivian's, though now the camera stays outside, watching them interact through a window. This shot looking into the window from the garden is unique in the film, but many shots from the interiors later use windows and open doors to invite the garden into the house and show the interdependency of the tropical plants and the living spaces carved out in their midst.

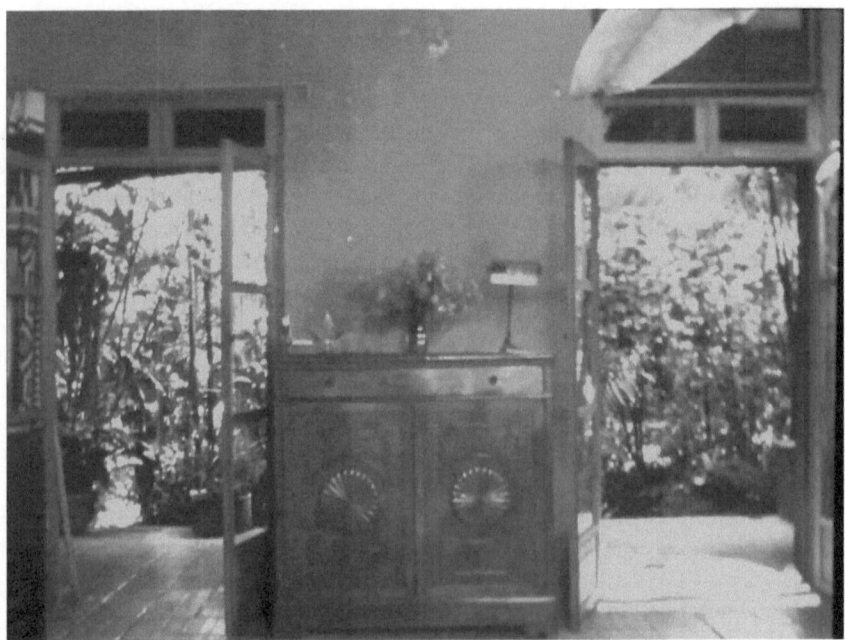

Figure 6.5 Through the window-like doors, the garden shines into even the interior spaces. Rosalind Nashashibi, *Vivian's Garden*, 2017, frame enlargement.

Starting with the caretaker is one of the ways the film emphasizes the presence of the support staff of Guatemalan women and men that make the artists' lives possible, especially since Elisabeth is confined to a wheelchair. Later in the film Elisabeth is shown paying Juan, the younger of the two workers, in bills taken from her purse, though we don't know if this cash exchange is salary or reimbursement for items purchased.

Only after this initial sequence do Vivian's activities become central; she rearranges a covering on a couch to accommodate one of the four dogs that live with the artists, as the title of the film appears onscreen. Next a short montage of three different dogs resting on the furniture or floor suggests a temporality that is no longer specific to one day, but rather implies habitual and repetitive actions.

After this canine interlude the film shows Vivian walking outside in the thick growth and deep shade of her tropical garden. She speaks of fear of the snakes as she reaches into the bushes, and in another shot, waves a large dead palm frond, as if demonstrating how like a giant brush it is. Yet the gesture is offered without any explanation that would directly link it to her art practice, instead hovering rhythmically at this stage of the film to show her spritely interaction with her garden.

A return to the garden's gate interrupts, emphasizing how the caretaker negotiates with the world outside. This time the weight of the large iron door is emphasized. An accented line of light filigreed by the ironwork is visible at its upper edges. When Don Tomás opens a slat in the door, more iron grillwork is seen framing a waiting newcomer, Juan. This is how the film describes the closed-off and sheltering aspects of the enclave, using visual exposition, sans voiced narration.

Another scene shows the Guatemalan women cleaning and preparing food, then one serving it to Elisabeth and Vivian. The two men, Don Thomás and Juan, are shown at another table, eating separately. With these shots, the film suggests the postcolonial implications of Europeans' expatriation in Central America. The relationship between these hired workers and the artists is central to the imagery, while their separate roles are sharply delineated.

When asked by Mulvey in the interview cited above about the visual rendering of these relationships between the artists and the Guatemalan support staff in the absence of more direct geopolitical commentary on economic exploitation, the filmmaker said she wished to leave that particular analysis implicit rather than overt. In other words, this is a film that uses visual representation to make its points, rather than more intrusive commentary; yet Nashashibi through visual means wanted her viewers to see how the women who clean and cook make the lives of these artists possible, even as they occupy this space as workers not as equals. As background to this visual inscription of the workers, let me note that in 2019 the United Nation's Women's Fund for Gender Equality took up the cause of domestic labor in Guatemala, "pushing for better income, health insurance, safer working conditions; and helping exploited domestic workers take their cases to the courts."[10] Every indication in the film suggests that the workers are not among the most exploited that these reforms were trying to address; for example, we see them eating lunch, presumedly in the middle of their workday, while one of the complaints the UN is trying to address is workers who are forced to work twelve hour days with no food breaks at all.[11] The relative brevity of the workers' appearance in the film and the lack of direct address to the workers' lives in the film should not be read as a simple or unfortunate consequence of the artistic focus of the film. Instead, consider how in modern film art a single shot can speak volumes to social context and invite an astute audience to analyze this politically.[12] Audiences are invited to think through the implications alongside the very insistence the film makes throughout on a strategy of showing, implying, and interstitial suggestion, that is, an aesthetics that trusts the spectator to think.

From here the film moves on to its characteristic flow of images: a large leaf shining in the interior of the house, the paintings of various sizes hung on walls, shadows of leaves on an outside wall. A conversation between the two artists floats over this montage. Fragmented phrases are offered by Vivian's lilting voice, some poignant, such as "we were like prisoners." This is a reference to a story that is not possible to fully decipher from the film alone, that of an incident in which a neighbor, who has since been jailed, threatened them if they ventured outside. Then over a series of garden images, we hear "he was really aggressive." This fragmented narrative challenges us to imagine the whole story, which remains elusive. Yet the salient elements, the emotional remembering becomes more poetic in this form. "I close the house tight every night, and that's what the dogs are for," Vivian says.

In the interview with Mulvey, Nashashibi fills in background that much of the discussion between the artists about fear is linked to rival drug gangs that had been fighting in the area, and that the fragments quoted above refer to a menacing criminal neighbor who had made it impossible for them to leave their enclave. Nashashibi speaks to the doubleness of the garden as both refuge and danger. Similarly, healing is offered as a gift by the caretaker in another sequence in the film. We see Vivian lying in her bed, asleep, intercut with images of Don Tomás putting banana leaves over a skylight on the roof of her house. The large leaves protect her slumber, but at first this sequence seems merely odd. In fact, in the context of the rest of the film, one understands this as a statement about the garden as shelter. If more classical gardens can yield their flowers and herbs to enhance our lives in interiors, as well as our interior lives, this garden can be helped to do the same by this gesture of protecting sleep. The film was shot in the spring, so the garden is not yet in full flower; rather it is richly laced with a multitude of shades of green, in a wide variety of shapes. Benchoam in his description of his visit to the garden lists an impressive range of plants and trees: "Areca palms, Royal palms, fishtail palms, bougainvillea, pine, ficus, eucalyptus, gavileas, monster deliciosos, Spanish asukas, birds of paradise, bamboo, macadamias, lemon, lime, sabal palms, justicia gardenias, tangerine, mangos, papayas, avocado, maiden hair ferns, Boston ferns, cinnamon ferns, Australian ferns, ostrich ferns, anthem ferns, and coffee plants."[13] As the filmmaker tells us, "The garden is a created universe, not a wild place. It feels like it's a jungle, but it is a garden that Vivian has grown over thirty years." Nashashibi cites Michael Taussig's book, *Shamanism, Colonialism, and the Wild Man: A Study in Terror and Healing*, as coloring her thinking about expressing the parallel terror and healing that this garden encompasses.

Let us recall elements of Guatemalan history in the post-World War II period that run parallel to Elisabeth's and Vivian's struggles to survive Nazism, exile, and their transatlantic searches for refuge. The national trauma of Guatemala has as underlying causes struggles over land tenure, coupled with repression of the underclass and disenfranchised. The legacy of colonialism, with its oppression of Mayas, descendants of enslaved peoples from Africa, and mestizos (terms used in the Guatemalan literature) was continued in Guatemala by corrupt regimes and international corporations. However, in the immediate aftermath of World War II, Guatemala had ten years of hopeful emergence from colonial and neo-colonial domination. In 1944, Guatemala had its own October Revolution that led to a pe-riod of ten years of democratically elected government, constitutional rule, and large-scale social reforms, granting voting rights to women and the poor, and end-ing forced labor. This democratic change culminated in the agrarian reform law of April 1952. As chronicled in Stephen Schlesinger's and Stephen Kinzer's *Bit-ter Fruit: The Story of the American Coup in Guatemala*, the Unites States under President Eisenhower initiated CIA covert operations against the elected govern-ment of Jacobo Arbenz.[14] This led to the 1954 coup, with one of the driving forces being the influence of the United Fruit Company in the Eisenhower administration, a conglomerate which owned all of Guatemala's banana production, as well as Guatemala's telephone and telegraph system, and most of its railroads. Following the coup, the US-installed regime of General Miguel Ydígoras Fuentes repressed all opposition, ruling corruptly. In 1963, the United States supported a second coup in Guatemala, staged to prevent Juan José Arévalo from returning to the country and running for re-election. Over three decades Guatemalan security forces under right-wing regimes killed an estimated forty thousand people. The killing radical-ized opposition to the government. The end of Guatemala's civil war in 1996 did not mean an end to violence in the country. While agrarian activists hoped that peace negotiations between the military and the guerrillas would resolve Guate-mala's land problem, these issues remain.

Further, violence in contemporary Guatemala remains widespread, often at the local level, such as the threats to Vivian's and Elisabeth's well-being alluded to in the film. What this history suggests is that the microcosm of Vivian's garden enclave stands in the shadows of much violence and repression; if the artists and their employees live there in peace and mutual shelter, it also invites us to see all the ironic historical resonances of this space.

The interior scene that lasts longest away from the garden occurs in Vivian's bedroom as she consults her mother about which of her clothes she should pack for an upcoming trip to Greece. Unlike the other dialogue in the film that is in English, this conversation is in German. Obliquely, this scene both establishes the German-speaking past of the family and Elisabeth's early career as a textile designer in a reference to fabric she has given her daughter. Yet, Elisabeth responds to some gar-ments that were made from cloth Vivian tells her were a gift from her by saying that she doesn't remember. Vivian handles the garments as if she is trying to give back to her mother the history woven into that cloth, just as Vivian delicately solicits her mother's advice. The camera often pans between the two women, and these pans catch glimpses of one of Vivian's large paintings on paper in tones of orange and sienna. The subtext of this interlude of sustained, non-fragmented exchange in the interior space is the worry each has about the pending departure. By the end of the scene, each seems to want to assure the other that all will be well in Vivian's absence, the mood lightened by banter about how the dog Sophie, who has been lounging on the bed the whole time, will take care of Elisabeth, while Elisabeth takes care of Sophie.

Another of the film's montage sequences that includes shots of both artwork and collected objects, as well as images of Vivian wheeling her mother through the garden, follows. Over the reprise of Elisabeth working on a collage by perusing one of the magazines, Vivian says in a voice-over fragment, "I cherish every moment with my mother." The most sustained shot of Elisabeth's collages lying on her workplace tabletop occurs in this montage sequence. The film's montage thus emphasizes that their art and lives are intertwined; they collaborate with their shared space, and though they work in different media their parallel endeavors are marked by their shared mode of being.

Next, Vivian in voice-over speaks of her fear when it rains, making the film's only reference to Hurricane Stan obliquely with a "because I remember" that trails off. "The strong winds, with all the trees around" is the voice-over that accompanies some very dark scenes shot at night, with spot lighting on certain parts of the frame showing the vegetation. The film returns to the exterior, to vegetation shots, emphasizing the dappled light filtering through the tropical trees.

A return to the garden in daylight marks the transition to shots in one of Vivian's two studios, as Juan helps her move a very large painting. Some shots show fragments of other paintings. Then a fragment of an animated film intrudes on a monitor, a trace perhaps of certain figurative cartoon-like elements that have recently been incorporated in Vivian's works. The film ends with Vivian preparing a bucket of pigment, including some mysterious sticks that she adds, then taking a canvas outside to work on in the garden.

Elisabeth Wild and Vivian Suter, as well as Marie Menken, are artists whose appreciation came with temporal lag. Menken has been far more recognized after her death than she was during her lifetime when exhibition of her work was far more limited. Vivian actively retreated from the art world only to be rediscovered recently. Nashashibi, on the other hand, is gaining recognition as a filmmaker, including being awarded the Turner Prize in 2017.

I have traced in this chapter their shared belief in abstracted visual expression, the poetic power of fragments and gestures for which the garden serves as sustaining space. For these artists the garden is a reflector of light, the shaper of shadows, the spaces in which bodies and cameras move, reframing and creating montages. These films are small, botanically inspired compositions, almost private in their mode, yet luckily finding their place in the history of women's artmaking. The sensitive, self-contained expression of power in these films rests on an implicit feminism for which the garden serves as an inspiration and support.

Notes

1 Mary Kocol, "The Garden in Early Art Photography."
2 Brakhage, "Marie Menken," *Film at Wit's End*, 33–48.
3 The New York Botanical Garden's archives are searchable online. See The Barnaby Catalogue for the material used to research his contributions.
4 Ragona, "Swing and Sway: Marie Menken's Filmic Events," 33.
5 Marx, *The Machine in the Garden*, 343.
6 Turim, *Abstraction in Avant-Garde Films*, 77.
7 Benchoam, in a talk given at *The Power Plant*, Oct. 20, 2018.
8 This story was told by Benchoam in his talk at *The Power Plant*, Oct. 20, 2018.
9 Interview of Rosalind Nashashibi by Laura Mulvey at the Horse Hospital Art Gallery, September 12, 2018. Elizabeth Wild died at age 98 in 2020. https://www.artnews.com/art-news/news/elisabeth-wild-dead-98-1202677852/
10 https://www.unwomen.org/en/news/stories/2019/3/feature-story-social-protection-to-domestic-workers-in-guatemala
11 Ibid.
12 For example, consider the single, high-angle shot of Japanese soldiers marching through the narrow street near the enclave in which the lovers seclude themselves in Nagisa Oshima's *In the Realm of the Senses*, 1976. As Stephen Heath first pointed out, and which I discuss in my book *The Films of Oshima Nagisa: Images of a Japanese Iconoclast* (127) the brevity of this intrusion of the historical marker of Japan's imperialist conquest of its Asian neighbors can be seen as more powerful both artistically and politically than a more common means of establishing social context in fictional film. It is a device that is accentuated by its inscription in the context of the modernist form of the entire film. In the documentary tradition in which voice is often used to drive home such issues, the adoption of the modernist documentary form tends to ask viewers to draw similar insights themselves from visual cues and juxtapositions of shots.
13 Benchoam, in a talk given at The Power Plant, Oct. 20, 2018.
14 Note that these issues are discussed brilliantly by Schlesinger and Kinzer in *Bitter Fruit.*

References

Allen, Jamie M. and Sarah Anne McNear. *The Photographer in the Garden*. New York: Aperture, 2018.

Barnaby Catalogue of the New York Botanical Garden. https://sweetgum.nybg.org/science/projects/barneby/narratives/?irn=1323

Benchoam, Stefan. Talk given at *The Power Plant*, Oct 20, 2018. https://www.youtube.com/watch?v=npvx0Y01UpI

Brakhage, Stan. *Film Biographies*. Berkeley: Turtle Island, 1977.

Brakhage, Stan. *Film at Wit's End: Eight Avant-Garde Filmmakers*. Kingston, NY: Documentext/McPherson and Company, 1989.

Crase, Douglas. *Both: A Portrait in Two Parts*. New York: Pantheon, 2004.

Kocol, Mary. "The Garden in Early Art Photography," *Gardens Illustrated* (2010).

MacDonald, Scott. *The Garden in the Machine: A Field Guide to Independent Films*. Berkeley: University of California Press, 2001.

Marx, Leo. *The Machine in the Garden: Technology and the Pastoral Ideal in America*. New York: Oxford University Press, 1976.

Nashashibi, Rosalind. Interview of Rosalind Nashashibi by Laura Mulvey at *The Horse Hospital Art Gallery*, September 12, 2018. https://www.youtube.com/watch?v=nxZLTe11o4g

Ragona, Melissa. "Swing and Sway: Marie Menken's Filmic Events," *Women's Experimental Cinema: Critical Frameworks*, Robin Blaertz, 2007: 20–44. https://doi.org/10.1215/9780822392088-002

Schlesinger, Stephen C. and Stephen Kinzer. *Bitter Fruit: The Story of the American Coup in Guatemala*. Cambridge, MA: Harvard University, David Rockefeller Center for Latin American Studies, 2005.

Suárez, Juan A. *Myth, Matter, Queerness: The Cinema of Willard Maas, Marie Menken, and the Gryphon Group 1943–1969.* Grey Room 2009; (36): 58–87. doi: https://doi.org/10.1162/grey.2009.1.36.58

Taussig, Michael T. *Shamanism, Colonialism, and the Wild Man: A Study in Terror and Healing*. Chicago: University of Chicago Press, 1991.

Turim, Maureen. *Abstraction in Avant-Garde Films*. Ann Arbor, MI: UMI Research Press, 1995.

Turim, Maureen. *The Films of Oshima Nagisa: Images of a Japanese Iconoclast,* Berkeley: University of California Press, 1998.

United Nation Social Protection to Domestic Workers in Guatemala. https://www.unwomen.org/en/news/stories/2019/3/feature-story-social-protection-to-domestic-workers-in-guatemala

7 Virginia Woolf and Vanessa Bell at Kew Gardens

Elise L. Smith

Virginia Woolf's *Kew Gardens*, written in 1917 during the war, was first published in May 1919 as a small hand-printed and hand-bound book. Both this booklet and the 1927 edition included woodcuts by her sister Vanessa Bell and were printed by the Hogarth Press, established by Woolf and her husband Leonard.[1] The innovative partnership between the two sisters, and between text and image in the two editions, was grounded in the living gardens that inspired them—their own gardens at Asheham House, Monk's House, and Charleston Farmhouse, as well as the eponymous botanical garden at Kew. The sisters' private, domestic plantings aimed at cottage-garden effects, unlike the long-established botanical institution with its conventional goals focused on conservation. Both approaches to garden-making—loose and intuitive, on the one hand, and well-ordered, on the other—have parallels in the stylistic innovations of *Kew Gardens*, but the booklets are ultimately far more radical in their modernist collaboration than the designs or plantings of these actual gardens.

The story is experimental in many ways, providing the reader with a remarkable juxtaposition of the human and natural worlds: the fragmented conversations of the couples strolling through Kew join with the shifting light effects and burgeoning, buzzing activity in the gardens around them. But the impetus for the project was not a modernist garden, and in fact there was not much interest in Britain in such an approach to landscape design until the 1930s with Christopher Tunnard, who urged his colleagues in 1937 "to create by experiment and invention new forms which are significant of the age from which they spring."[2] Twenty years earlier, Woolf and Bell were among the creative women who had been doing just that in their respective art forms, though the style of their own gardens remained in the Arts and Crafts tradition, as we will see, and Kew itself was founded on ideals of scientific order. Begun in 1759 by Princess Augusta, mother of George III, as a nine-acre botanic garden, Kew came to be seen by many as "a silly retreat from royal responsibility," an "unnatural, over-managed, elitist and faddish" folly.[3] But by 1844 William Hooker, who had been named director four years earlier, felt confident in stating that "there scarcely exists a garden or a country however remote, which has not already felt the benefit of this establishment."[4] Its role as a unifying force in the Empire, both economic and political as well as biological, continued to be touted, and a Treasury Committee in 1900 described Kew as having "a distinctly Imperial character."[5]

DOI: 10.4324/9781003381549-8

By the early twentieth century the Royal Botanic Gardens had thus come to serve as a powerful sign of scientific progress and colonial expansion.[6] Although its emphasis on categorizing the vastness of the botanical world is nowhere in evidence in Woolf's atmospheric story, she does allude, seemingly randomly, to some specimens brought from the colonies: orchids, palms, a red waterlily.[7] But her love of the gardens was neither political nor economic, nor was it based on a scientific interest in botany. She could see Kew through the window of the house in Richmond where she and Leonard moved in 1915, and she took pleasure in the effect of passing seasons on the gardens there. She often mentions walks in Kew in her diary and was clearly interested in the variety of plants and insects: she writes on 25 January 1915, "After lunch, I met L. at the gates of Kew Gardens, & we walked back to Richmond through the Gardens, which are now one feels teeming with buds and bulbs, though not a spike shows"; and on 12 March 1918, "I sat by choice on a seat in the shade at Kew; I saw two Heath butterflies; willows, crocuses, squills all in bud & blossom."[8] She was even aware of which days had the reduced entrance fees, which determined when she would visit.[9] Later, returning to London in 1924 after a decade in Surrey, she continued to explore Kew; one such occasion was in 1926 with Vita Sackville-West, shortly before her reference to the same gardens in *Orlando* (1928): "dreaming of more than can rightly be said," she imagines spring bulbs "thrust into the earth."[10] In her visits she seems to have been interested more in the small details, the "sensation," rather than the scientific rationale for a botanical garden or the impressive vistas to be found there.[11]

These poetic particularities of organic growth were also at the heart of Woolf's delight in the "fertility & wildness" of her own garden.[12] She and Leonard spent weekends and holidays in East Sussex, first at Asheham House in East Sussex before their move to Monk's House in 1919, and Vanessa Bell moved close by to Charleston Farmhouse in 1916, with Duncan Grant. Both households spent as much time as possible outside, and the gardens were seen as extensions of the interior spaces. Bell was a painterly gardener, eager for color ("a dithering blaze of flowers and butterflies and apples," as she later described it), but she was also concerned about practical matters: she repeatedly laments in her letters about the incessant problem with weeds.[13] Unlike her sister, Woolf had little interest in the actual work of gardening, although she did occasionally make references in her diary: "Did the garden path all the afternoon. Planted some wall flowers, daisies, foxgloves."[14] In general, though, the garden—especially later at Monk House—was largely Leonard's concern, as she indicates in a 1925 letter: "I offer my admiration, but am seldom allowed an active part."[15]

But Woolf was deeply attuned to the rhythms of nature, as Christina Alt and Bonnie Kime Scott have both recently explored. In 1917, the year that she wrote this story, she was coming out of a prolonged period of mental illness and coping with the trauma of the Great War. Her diary entries tended to be shorter than usual, often related to mundane domestic issues but also to her observations of nature, as if the listing of plants and brief descriptions of animal behavior could provide some measure of stability, if not solace: "Saw 3 perfect peacock butterflies, 1 silver washed frit[illary]; besides innumerable blues feeding on dung"; "Ladies Bedstraw,

Round-headed Rampion, Thyme, Marjoram. Saw a grey looking hawk—not the usual red-brown one."[16] In *Virginia Woolf and the Study of Nature*, Alt suggests that Woolf's writing, both personal and published, indicates she was drawn less to a taxonomic approach to nature, as at Kew, than to ethology, or the close study of animal behavior in its larger context.[17] We can apply this insight to *Kew Gardens*, which I want to consider not as an isolated text, as it usually is, but instead more holistically with text and images as partners in meaning. Both the 1919 and 1927 editions, then, will serve as reminders of the conflicted views of the garden characteristic of Woolf and Bell, as well as other women at the time.[18] As a site of both professional opportunity and personal interest, the garden for these sisters was demanding and claustrophobic as well as invigorating and restorative.

Kew Gardens as a text has received a great deal of critical attention, beginning almost immediately after publication with a positive anonymous review by Harold Child in the *Times Literary Supplement* (29 May 1919).[19] Although he summarizes the narrative as being "about Kew Gardens and a snail and some stupid people," he recognizes that the subject is essentially unimportant. Kew as the context was a mere pretext, since the gardens themselves "are neither something nor nothing; neither formal nor wild; neither old nor new; neither urban nor rural; neither popular nor choice." He goes on to wonder, "What are Mrs Woolf and Mrs Bell going to find in Kew Gardens worth writing about, and engraving on wood . . .?"[20] Their creative detachment from the physical reality of Kew, despite their own familiarity with that reality, is in fact what enables the modernist experimentation of the booklet.

Critical responses have vacillated between reading the story as pure formal experimentation and as having a political subtext. E. M. Forster emphasizes the lack of a moral center in his review in the *Daily News* (31 July 1919). His descriptions—"the surfaces of things," the world of the Eye," "vision unalloyed"— situate this story in the aesthetic realm of Impressionism, as does his emphasis on its formlessness: "It aims deliberately at aimlessness, at long loose sentences that sway and meander; it is opposed to tensity and intensity, and willingly reveals the yawn and gape."[21] More recently, John Oakland rejects the view that Woolf was responding to the meaninglessness of life, instead proposing that she was conveying "a harmonious, organic optimism" and that her tone here is "joyful rather than despairing."[22] Other contemporary critics often point out political undercurrents: Alice Staveley sets this "unquestionably formalist" story in the context of what she sees as Woolf's feminist narratology, and Jane Goldman states that "It is impossible to understand the formalist aspects of modernist aesthetics as occurring in a political vacuum," relating Woolf's style to the "shocking colourism of both Post-Impressionist and suffragist art."[23] The Woolfs must have known of the attacks in 1913 on the Tea Pavilion and Orchid Houses at Kew by two Suffragettes, who were looking for popular tourist sites that would generate as much publicity as possible for their cause; Kew certainly fitted the bill, since it had attracted 3.8 million visitors in 1912.[24] Reading the story from a different political angle, Shelley Saguaro sees it as "an exercise in experimental fiction" and an "evocation of an atmosphere" but also notes that Woolf was assuredly aware of Kew as a monument to "British botanical imperialism."[25]

I recognize how Woolf subverts the implicit imperial dominance of the garden by showing its fragility: the visitors attend only fitfully to its particularities, thoughts of the war take precedence at times, and the noise of the city impinges on nature in the conclusion. But I am also drawn to the aesthetic reading that has been proposed, beginning with the early reviews: the formlessness akin to Impressionism, the underlying structure suggesting a Post-Impressionist style that she would have been introduced to in Roger Fry's exhibition in 1910, if not before. This approach enables us to respond more readily to the booklets in their entire form, as a collaboration of text and image: as Child wrote in the first review, the book itself is "a work of art, made, 'created,' as we say, finished, foursquare; a thing of original and therefore strange beauty, with its own 'atmosphere,' its own vital force."[26] I, too, find particular importance in the material reality of the book itself, its tactile quality enhanced by the carefully selected papers, the hand-printing of the first edition, and the remarkable covers: Roger Fry's moody, marbled paintings for the 1919 edition have a very different effect than Bell's woodcut for the cover of the 1927 booklet.[27] Both, however, are non-narrative, thus denying readers any hint of character or plot development, as is more typical of covers for popular novels of the period; instead, we dwell on the decorative qualities, using a word favored by the Bloomsbury circle.

Bell described the story as "fascinating and a great success, I think," after reading it in manuscript form (despite Woolf's apology for it being "very bad now, and not worth printing"), and she asked her sister if she might do a print for it: "It would be fun to try, but you must tell me the size. It might not have very much to do with the text, but that wouldn't matter."[28] Woolf wrote in response, "I don't see that it matters whether it's about the story or not," and she suggested that the size of the prints should be about the same as Dora Carrington's illustrations for *Two Stories*.[29] Carrington's four woodcuts were actually the Woolfs' first collaborative effort in print—with each other and with an artist. The short book appeared in July 1917, just a few months after they bought a printing press and instruction manual from the Excelsior Printing Supply Company. Much of the tedious work of producing a book on their equipment was done by Virginia, since Leonard's hand tremor made typesetting and binding difficult, but she loved the process; as she wrote to a friend, "You can't think how exciting, soothing, ennobling and satisfying it is."[30] Woolf was particularly pleased by the use of woodcuts, mentioning in a letter to Carrington shortly after the publication, "we see that we must make a practice of always having pictures."[31] For *Kew Gardens*, Bell made only two woodcuts for the first edition, but she created a complete new set for the 1927 booklet, one for each page in addition to the cover. These two versions, then, have quite different effects as we consider the interaction of word and image.

But first, the story itself: it begins with a remarkable description of the small oval garden bed at Kew that almost serves as one of the protagonists. Woolf gives us anthropomorphic details ("heart shaped or tongue shaped leaves") that provide an initial indication of the sort of merging of human, animal, and botanical experience to come. The solid, almost heavy descriptors of the first sentence—"from the red blue or yellow gloom of the throat emerged a straight bar, rough with gold dust and slightly clubbed at the end"—turn ephemeral in the second, as the breeze stirs the petals and the light passes over the colorful flowers, "staining" the earth below. The fugitive light reveals unexpected details—a pebble, a snail, a raindrop, a leaf—before the color flashes up to distract the people walking by. There is nothing purposive about the men and women: they "straggle," their movements "curiously irregular" like the zig-zagging of the butterflies. Throughout the story Woolf draws parallels between the thoughts and actions of the visitors to Kew on that hot July day—four pairs wandering "carelessly"—and the natural environment, writ small in the oval border that they observe. In the first pairing, Simon remembers his marriage proposal to Lily fifteen years earlier and the circling dragonfly that he used as a predictor of her answer: "my love, my desire, were in the dragonfly." If it landed on a particular leaf then she would say yes, but it continued to circle, never settling, and here he is now, married to another woman. Woolf provides us with no explanation, no bridge between then and now, between memory and reality; the random nature of this scene is underscored by her description of the sunlight playing over their disappearing forms "in large trembling irregular patches" so that they appear "half transparent."

Woolf returns us to the microcosm of nature in the story's first interstice as she juxtaposes the slow, deliberate progress of the snail with the unpredictable gait of a "high stepping angular green insect."[32] The snail is clearly goal-oriented, though the goal is never definitively specified, and I agree with Oakland, who argues that this small creature "demonstrates purpose and achievement."[33] Could we, in fact, see the snail in its purposeful progress as akin to the typesetter of this booklet—Virginia Woolf herself? We might also find a parallel between the two insects and the next pair of garden visitors: Woolf contrasts the steady calm of the young man, characterized by a "look of stoical patience," with his elderly companion's "irresolute and pointless" gestures and "curiously uneven and shaky method of walking." The latter's almost incoherent words seem to signify the trauma of war as he refers haltingly to the Greek battle at Thessaly, to women in black, and to widows.[34] Woolf wrote to her sister about her own sense of numbness and confusion at the end of the war, precipitated by the uncertainties of the preceding years: on Armistice Day, referring to the noise of guns, sirens, whistling, and barking, she wrote, "its all done in such an intermittent kind of way that its not in the least impressive—only unsettling," and a few days later she mentioned her "dazed discontented aimless feeling."[35] She would seem to find common ground with the "irresolute" and "shaky" old man in this *Kew Garden* pairing, who apparently finds no solace in the beautiful gardens.

Two women then appear, confused by the old man's bizarre actions. Their own conversation, heard in meaningless bits and pieces, takes on the impressionist atmosphere of shifting light patterns described earlier—though now the "pattern of falling words" lacks any solid communicative reality, unlike the flowers which are "standing cool, firm, and upright in the earth." The fourth and final pair of garden visitors speak in "toneless and monotonous voices" with long pauses between their inconsequential comments; caught in that moment of stasis on the cusp of maturity, like a bud or chrysalis, they are

> both in the prime of youth, or even in that season which precedes the prime of youth, the season before the smooth pink folds of the flower have burst their gummy case, when the wings of the butterfly though fully grown are motionless in the sun.

But Woolf assures us, using the image of a bumbling bee, that their "short insignificant words also expressed something, words with short wings for their heavy body of meaning, inadequate to carry them far and thus alighting awkwardly upon the very common objects that surrounded them . . . " After the incoherence and uncertainty of the preceding paragraphs, she tantalizes us with the comforting reality of tea-time—"Come along, Trissie; its [sic] time we had our tea"—but we are still not allowed to leave the fugitive trivialities of these meandering humans. Trissie is "looking vaguely round her and letting herself be drawn on down the grass path, trailing her parasol, turning her head this way and that way, forgetting her tea, wishing to go down there and then down there . . .."[36]

In the last paragraph, the heat of the day shimmers and fractures any sense of purpose. Impressionist formlessness turns the garden and its visitors into a kind of mirage:

Thus one couple after another with much the same irregular and aimless movement passed the flower-bed and were enveloped in layer after layer of green blue vapour, in which at first their bodies had substance and a dash of colour, but later both substance and colour dissolved in the green blue atmosphere.

The strong colors of the flowers at the beginning of the story are now muted, there are long pauses between the thrush's hops, and the white butterflies dance above the flowers, "making with their white shifting flakes the outline of a shattered marble column." In the glare of July heat, the humans "wavered and sought shade beneath the trees, dissolving like drops of water in the yellow and green atmosphere, staining it faintly with red and blue." Only their voices remain to rise above the motionless forms, "wordless voices, breaking the silence"—but no, not silence. In the last sentence of the story Woolf returns us from this languid pastoral scene to the urban noise and action of London:

all the time the motor omnibuses were turning their wheels and changing their gear; like a vast nest of Chinese boxes all of wrought steel turning ceaselessly one within another the city murmured; on the top of which the voices cried aloud, and the petals of myriads of flowers flashed their colours into the air.

The nesting boxes serve as an image of the complex ordering of nature, with all components interrelated: nature and the metropolis, humans and the smallest creatures, form and formlessness.[37]

Woolf's commitment to the formalism of Roger Fry, with his ideas about order and unity, must be seen through the lens of her "life-long preoccupation with dissolution and nothingness," as Saguaro suggests, though for Woolf dissolution was more about creation than negation.[38] Impressionist and Post-Impressionist art, familiar to the sisters as both object and theory, provide a useful aesthetic counterpart to consider the collaboration of text and image in the two editions of *Kew Gardens*.

Woolf herself described the story as "a case of atmosphere" in a 1918 letter to her sister, using language reminiscent of Impressionism, and Forster recognized the connection in his review: the flowers serve a purely visual rather than allegorical function, as they "cause us to see men also as petals or coloured blobs that loom and dissolve in the green blue atmosphere of Kew."[39] The stylistic effects of the text do indeed have much in common with a work like Berthe Morisot's *Butterfly Hunt* (1874, Musée d'Orsay), which is characterized by loose, fragmented, seemingly random brushstrokes, by the fleeting perception caught by a "naïve" eye, and by the dissolution of form through light and atmospheric conditions. As Monet wrote,

> When you go out to paint try to forget what object you have before you—a tree, a house, a field or whatever. Merely think, here is a little square of blue, here an oblong of pink, here a streak of yellow, and paint it just as it looks to you, the exact colour and shape, until it emerges as your own naive impression of the scene before you.[40]

Woolf would certainly have had opportunities to see Impressionist art: this Morisot was included in an exhibition at the Grafton Gallery in 1905 of 315 Impressionist paintings, and the Woolfs also traveled to Paris in 1907 with Vanessa and Clive Bell.

But the atmospheric quality of *Kew Gardens* is also undergirded by a formal structure: in its most simplistic distillation, four pairs of garden visitors, separated in two interstices by the snail's enigmatic appearance, and framed by the introductory and concluding paragraphs that focus, respectively, on the garden and the city.[41] It is as if Impressionism takes on new strength by a Post-Impressionist spatial compression and formal rigor. In Woolf's biography of Roger Fry (1940), she quotes from his introduction to the Second Post-Impressionist Exhibition at the Grafton Galleries in 1912 (in which several of Bell's paintings were included): Fry writes that these artists "aim not at illusion but at reality," adding that

> they wish to make images which by the clearness of their logical structure, and by their closely-knit unity of texture, shall appeal to our disinterested and contemplative imagination with something of the same vividness as the things of actual life appeal to our practical activities.[42]

Fry's concern about structural clarity is a central element of his aesthetics, one that was also important to Woolf. Her description of a childhood memory in St. Ives from her autobiographical essay "A Sketch of the Past" (1939) is pertinent here:

> I was looking at the flower bed by the front door; 'That is the whole', I said. I was looking at a plant with a spread of leaves; and it seemed suddenly plain

that the flower itself was a part of the earth; that a ring enclosed what was the flower; and that was the real flower; part earth; part flower.

Woolf goes on to describe the impact that this had on her later thinking, that "one is living all the time in relation to certain background rods or conceptions. Mine is that there is a pattern behind the cotton wool. And this conception affects me every day."[43] What she calls the "rods or conceptions," or underlying pattern, form a framework to support the looseness of our perceptions, the "cotton wool." Or in her alternative explanation, more suited to the context of *Kew Gardens,* the vertical of the flower links the earth with the natural world, enclosed in a unifying ring. Woolf thus provides us with a fruitful conjoining of the modernist styles of Impressionism and Post-Impressionism at work in her story.

Woolf's emphasis on both "sensation" and "rhythm" as key components of her aesthetics helps us understand the integration of her text with her sister's woodcuts in *Kew Gardens*.[44] These were also important elements in Impressionist and Post-Impressionist artistic theory. First, "sensation," or what the senses take in (the Impressionists' "naïve" eye, Cezanne's "logic of sensations"[45]), was oriented around both light and color, as evident in Woolf's language though less so, at least initially, in Bell's black-and-white prints.[46] "Rhythm" becomes paramount as a unifier in *Kew Gardens*: in Bell's strong patterning and the juxtapositions apparent in her flattened spatial planes, and in the pace of the visitors' meandering as well as the overall structure of the text. As Woolf wrote in a 1926 letter, "Style is a very simple matter; it is all rhythm," and she went on to explain that

A sight, an emotion, creates this wave in the mind, long before it makes words to fit it; and in writing (such is my present belief) one has to recapture this, and set this working (which has nothing apparently to do with words) and then, as it breaks and tumbles in the mind, it makes words to fit it.[47]

Thus sensations and feelings come first, and words must follow their lead.

Gillespie and Goldman both recognize the importance of Bell for her sister's aesthetic decisions—just as much as, if not more than, Roger Fry and Clive Bell.[48] The relationship of the two women had professional as well as emotional and familial resonance, and the significance of *Kew Gardens* is based on their mutual contributions. In Woolf's foreword to the catalog for an exhibition in 1930 of her sister's work, she began by praising Bell's use of light and color in paintings that were "suffused, lit up, caught in a sunny glow." But she suggested that the paintings were not just about "delicious sensations," for "There is something uncompromising about her art ... They give us an emotion. They offer a puzzle. And the puzzle is that while Mrs. Bell's pictures are immensely expressive, their expressiveness has no truck with words."[49]

It can be said, then, that Bell's woodcuts for both editions of *Kew Gardens* amplify our understanding of the emotional tone of the book without re-presenting Woolf's words in another medium. They might best be considered extrapolations rather than illustrations. Her prints for the 1919 edition, a frontispiece and endpiece, provide a strong, bold framework for the narrative.

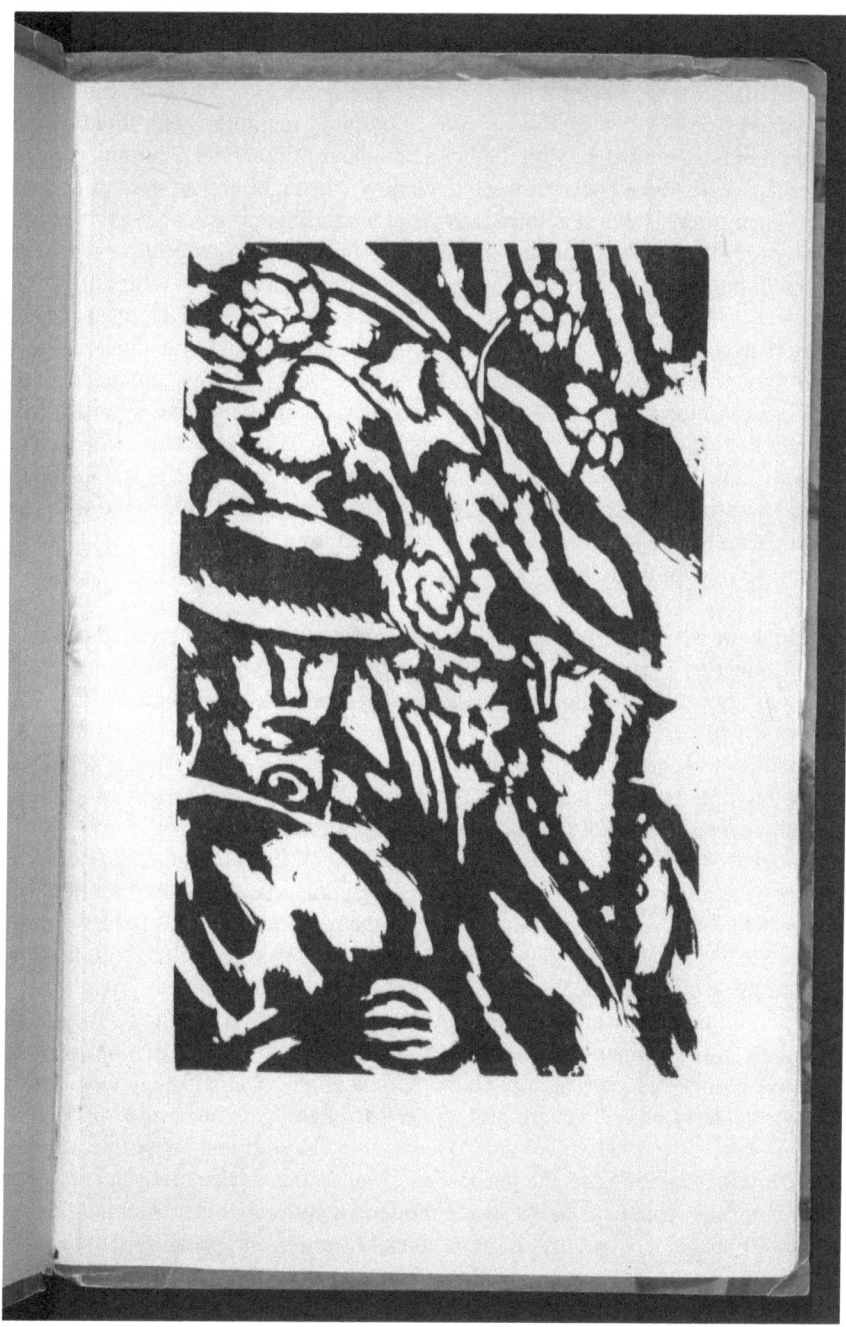

Figure 7.1 Vanessa Bell, woodcut frontispiece illustration in Virginia Woolf, *Kew Gardens*, 1919, p. 4. Photo courtesy of the British Library. © 2023 Artists Rights Society (ARS), New York/DACS, London.

The first woodcut (Figure 7.1) is related to the scene of the two lower middle-class women, but rather than being a direct visual commentary on the text it developed from an earlier painting of Vanessa's. In a letter to her sister Bell mentions her painting of *The Conversation* (1916, Courtauld Institute), with three women gathered in a tight cluster, behind them a bouquet of flowers in a curtained alcove or perhaps a flowerbed seen through a window: "It might almost but not quite do as an illustration."[50] Woolf admired her sister's painting, praising her as "a satirist, a conveyor of impressions about human life: a short story writer of great wit . . .."[51] As Woolf becomes painterly in her writing style, Bell becomes a writer of painted stories. But Woolf recognized that Bell created "stories" that were suggestive rather than literal stories to be felt rather than read. Although the composition of this first woodcut for *Kew Gardens* is quite different from the painting, with the characters reduced from three to two, the spatial compression and strong, almost abstract lines in the painting do reappear in the print. After Bell sent the design for the woodcut, Woolf responded that it was "a most successful piece; and just in the mood I wanted. As a piece of black and white it is extremely decorative—you see my language is already tainted."[52] She then asked for a second print, about half the size of the first, and wrote,

I think the book will be a great success—owing to you: and my vision comes out much as I had it, so I suppose, in spite of everything, God made our brains upon the same lines, only leaving out 2 or 3 pieces in mine.[53]

In addition to hinting at her personal traumas, Woolf's apologia also alludes to her belief, noted earlier, that words, the very tools of her own profession, can only partially convey the emotional core of "sensations."

In her first woodcut Bell used bold, heavy, black lines that create a dense flat pattern, seemingly abstract until the eye distinguishes these two women who merge with the flowers and foliage behind them. Although the harshness of the print style has none of the Impressionist fluidity of her sister's text, with no reference to the repeated reds, blues, and yellows mentioned in the early pages of the story, it still captures something of the blended figural and landscape forms. The diagonals of the woman's hat at the right continue into the lines of what seems to be a tree branch behind her, and we are left uncertain whether the flowers springing decoratively above her are part of her hat or the garden border.[54] There is a kind of vacancy to the two women's faces, hinted at with just a few interior lines, that conveys something of the purposeless, wandering, meandering quality of the visitors to Kew that day. Bell's strong modernist style, with its awkward disjunctions and spatial uncertainty, is more akin to the Post-Impressionist works that she would have seen—including the two-dimensional patterning of paintings by Gauguin, Van Gogh, Cezanne, and Matisse, among others—in Roger Fry's "Manet and the Post-Impressionists" exhibition in London in the winter of 1910 (notably related to Woolf's declaration that "on or about December 1910 human character changed").[55] Her own developing style led to the inclusion of four paintings by her in the Second Post-Impressionist Exhibition.[56] Although one might imagine

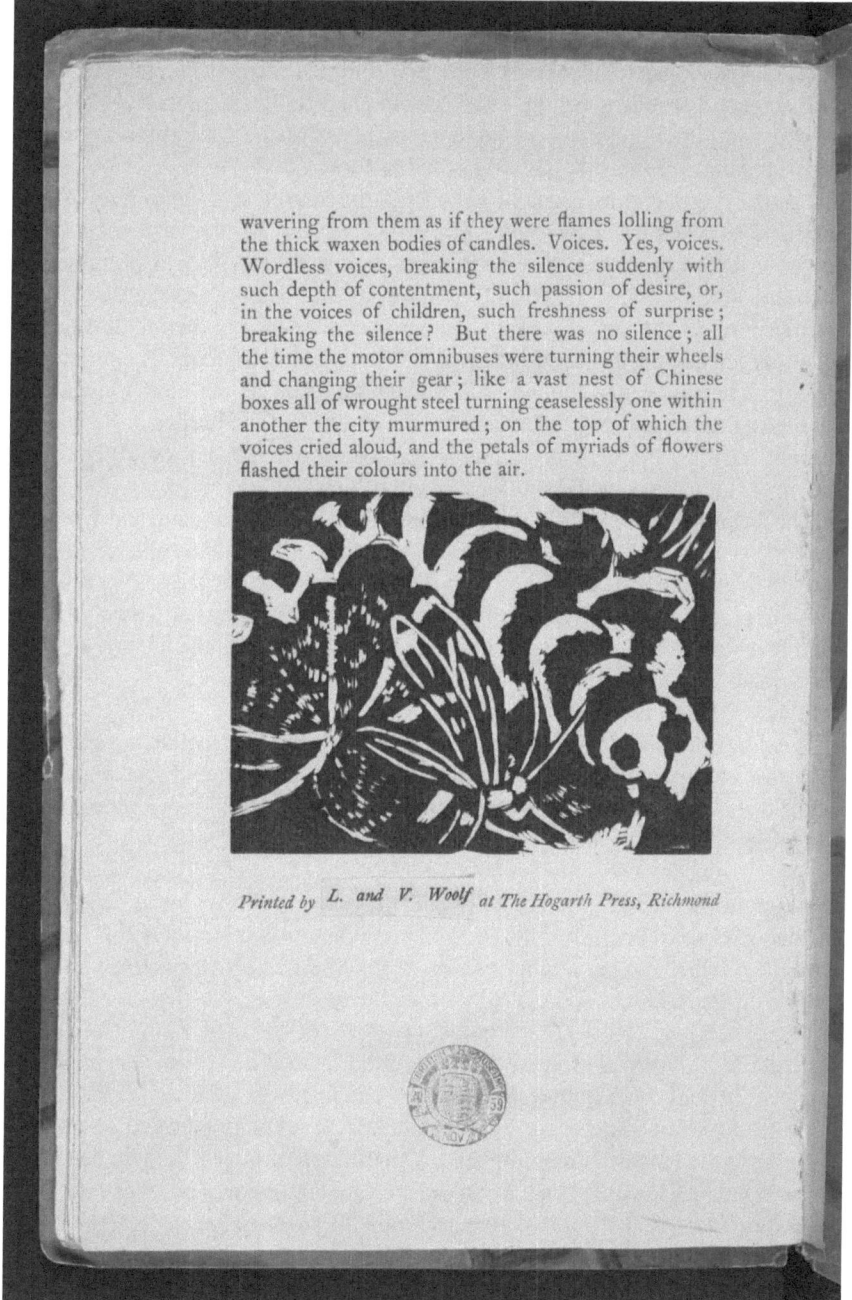

Figure 7.2 Vanessa Bell, woodcut end-piece illustration in Virginia Woolf, *Kew Gardens*, 1919, p. 10. Photo courtesy of the British Library. © 2023 Artists Rights Society (ARS), New York / DACS, London.

that painterly watercolors would be more suitable, in a way, to Woolf's shifting, impressionistic images, her modernist gaps and fissures, illogical and abrasive at times, make her sister's woodcuts in the first edition a particularly suitable pairing.

The second woodcut for the 1919 edition (Figure 7.2) is placed at the end of the text and now the flattened, overlapping shapes of the natural forms leave no room for human figures. A butterfly is in the foreground, the heavy woodcarving still suggesting some measure of delicacy through Bell's juxtaposition of white lines against black ink; curved behind it is a caterpillar, thickly segmented, with a flower or leaf to the left. Nature here seems dark and congested, as if pressed into the frame of the woodcut by the surrounding city, although otherwise Bell gives us no indication of the busy London streets that corral Kew. In fact, the inside back cover, directly opposite the last page of text with Bell's woodcut, is overlaid with paper that brings us into the world of the garden again. In the copy illustrated by Sorensen, the vibrant rosy colors and the outsize leaves and berries or flower, only partially visible, are shockingly contrasted with the small black-and-white print and must surely have been Woolf's deliberate choice.[57]

The technical challenges of printing the booklet were significant, and Woolf was involved at each stage, including setting up the type, binding the copies, and finally mailing them out.[58] She consulted with Bell throughout the process: experimenting with the proofs for her woodcuts ("not nearly black enough, but we only had our hand machine"), finding the right paper and ink, deciding on the number of copies and pricing, and figuring out how to split the costs:

> Wouldn't the simplest plan be to share the profits? first deducting our ex-
> penses of paper and postage and possible advertisements, and your expense,
> for wood blocks? We don't count our printing time; and I expect the artists
> time and the writers time were about equal in this case.[59]

But even in these final stages of producing the book, Woolf writes in her diary that she had read a final bound copy, an "evil task," and was disappointed: "The result is vague. It seems to me slight & short . . .."[60] She was also concerned about her sister's disappointment in the quality of the images:

> Nessa & I quarreled as nearly as we ever do quarrel now over the get up of
> Kew Gardens, both type & woodcuts; & she firmly refused to illustrate any
> more stories of mine under those conditions, & went so far as to doubt the
> value of the Hogarth Press altogether. An ordinary printer would do better in
> her opinion. This both stung & chilled me. Not that she was bitter or extreme;
> its her reason & control that give her blame its severity.[61]

They avoided more acrimony about the 1927 edition, since those 500 copies were printed commercially by Herbert Reiach, Ltd., in London.

Bell's prints for this later edition are very different in style and effect from the 1919 woodcuts.[62] Now described on the cover as "decorations" (rather than simply "woodcuts by Vanessa Bell," as in the first edition), they are similar to her painted

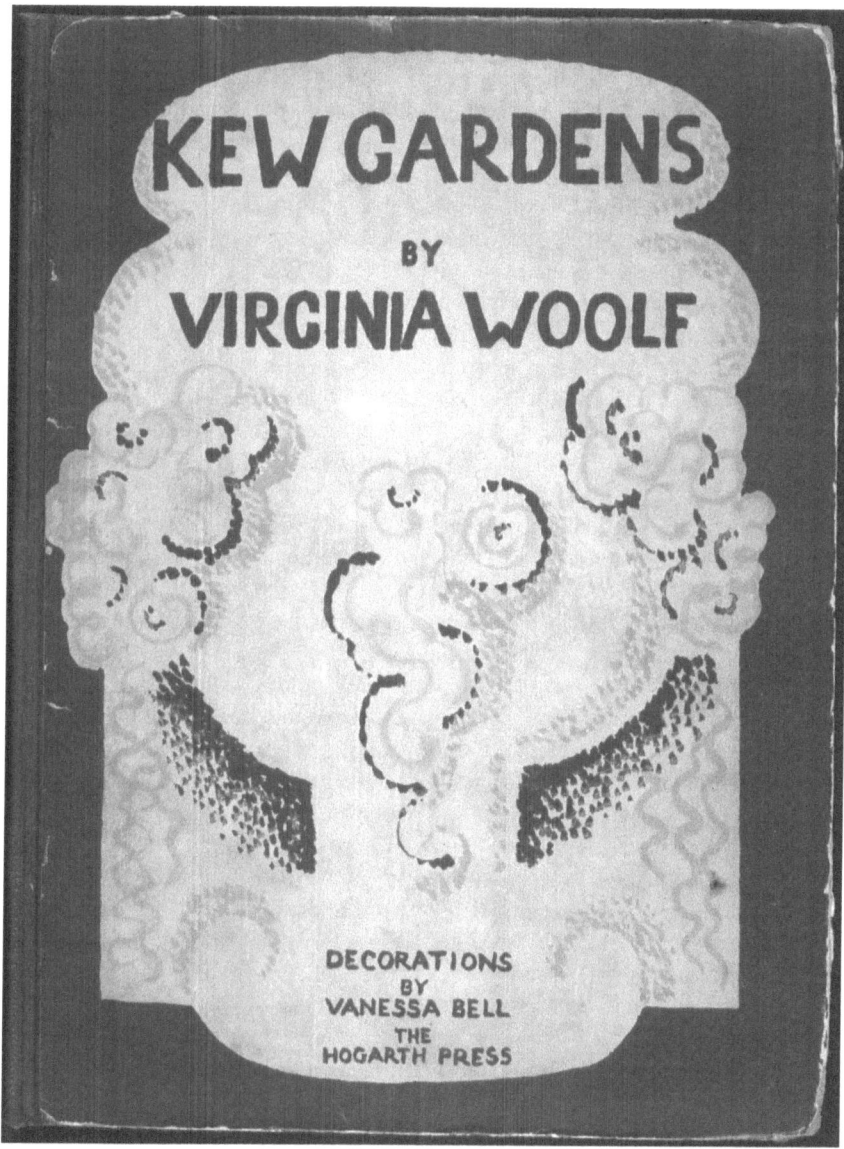

Figure 7.3 Vanessa Bell, cover for Virginia Woolf, Kew Gardens, 1927. Photo courtesy of
the British Library. © 2023 Artists Rights Society (ARS), New York / DACS,
London.

furniture, murals, and textiles for Charleston, her home in East Sussex, and her
work for the Omega Workshops, a design collective for decorative and applied arts
founded by Roger Fry in 1913.[63] This new set of woodcuts might also be considered a modernist version of William Blake's illuminated books, or even the kind of

marginal frames in medieval manuscripts. Frances Spalding, in her biography of Bell, proposes that they are successful because of their imprecision, thus not competing with the text, but they have a more prominent role than she implies.[64] Both Virginia and Leonard Woolf were careful in their instructions to the printer about the layout of the type in order to accommodate Bell's woodcuts, an indication that they considered her designs as co-partners with the text, not simply as peripheral.[65] The early order forms for the book also emphasize Bell's importance, as the designer as well as illustrator: "The whole book has been designed and each page is decorated by Mrs. Bell. The cover, printed in three colours, is also designed by her."[66] Thus Hermione Lee's description of the woodcuts as "visual underscoring" seems more accurate, signifying that they strengthen the impact of the text, along with Benjamin Harvey's suggestion that the prints exist in "the space between" illustration and decoration and in fact often amplify the meaning of Woolf's story in significant ways.[67] As we have seen, Woolf worried that her term "decoration" to describe Bell's first 1919 woodcut was "tainted," having the pejorative connotation of a minor art (even in Bloomsbury circles).[68] I doubt she would have objected to having "book illustration" substituted for "painting" in her later statement that "painting and writing have much to tell each other: they have much in common"; she then added, "The novelist, after all, wants to make us see."[69] Or a few years later, writing in more personal terms to her sister, she asked, "Do you think we have the same pair of eyes, only different spectacles?"[70]

Bell's approach was to wrap the text in frames, which in one sense makes the woodcuts seem secondary—that which one looks through, barely noticing—but at the same time makes them integral to the visual impact of each page, unlike more standard illustrations which are set apart. There is also considerable variety in the frames: some are delicate and graceful while others are looser or heavier. The more balanced and ordered frames often (though not consistently) align with moments of stasis in the text, and those that are strikingly asymmetrical usually "emphasize references to motion, tension, or confusion," as Gillespie notes.[71] Without being exact in an illustrative sense, Bell's cover design (Figure 7.3) hints at Kew Gardens with the three clusters of lush, billowing lines (reminiscent of foliage) and the overall shape of the frame itself, perhaps a variation on the curves of Kew's Palm House roof and entrance.[72]

The twenty-one pages of ornamented text inside the 1927 edition begin with a lively, curvilinear array of flowers, including the "heart-shaped or tongue-shaped leaves" described in the text (Figure 7.4), so we immediately experience the luxurious, burgeoning growth of the well-tended beds at Kew.[73] The woodcuts are indeed more decorative than illustrative, although Bell continues to give us references to flowers or vegetation, and occasionally even to the breeze that in the text ruffles the leaves and disrupts the sense of solid form; in Bell's woodcut, however, that breeze remains hard-edged in its looping curves (Figure 7.5), and only the half-seen blooms at the bottom reinforce the modernist fragmentation of the narrative. On another page, the movement of the dragonfly in the story of Simon and Lily is reduced to a spiral, endlessly in motion, never settling, although the decorative frame around the paragraph giving the commentary of Simon's wife—"Doesn't one always think of the past, in a garden with men and women lying under the trees?"—is more explicit. Here an ornamental tree rises at the left, its foliage mingling with the text, and the "ghosts lying under the trees" are indicated, surprisingly, by three particularly strong, straight, black lines.

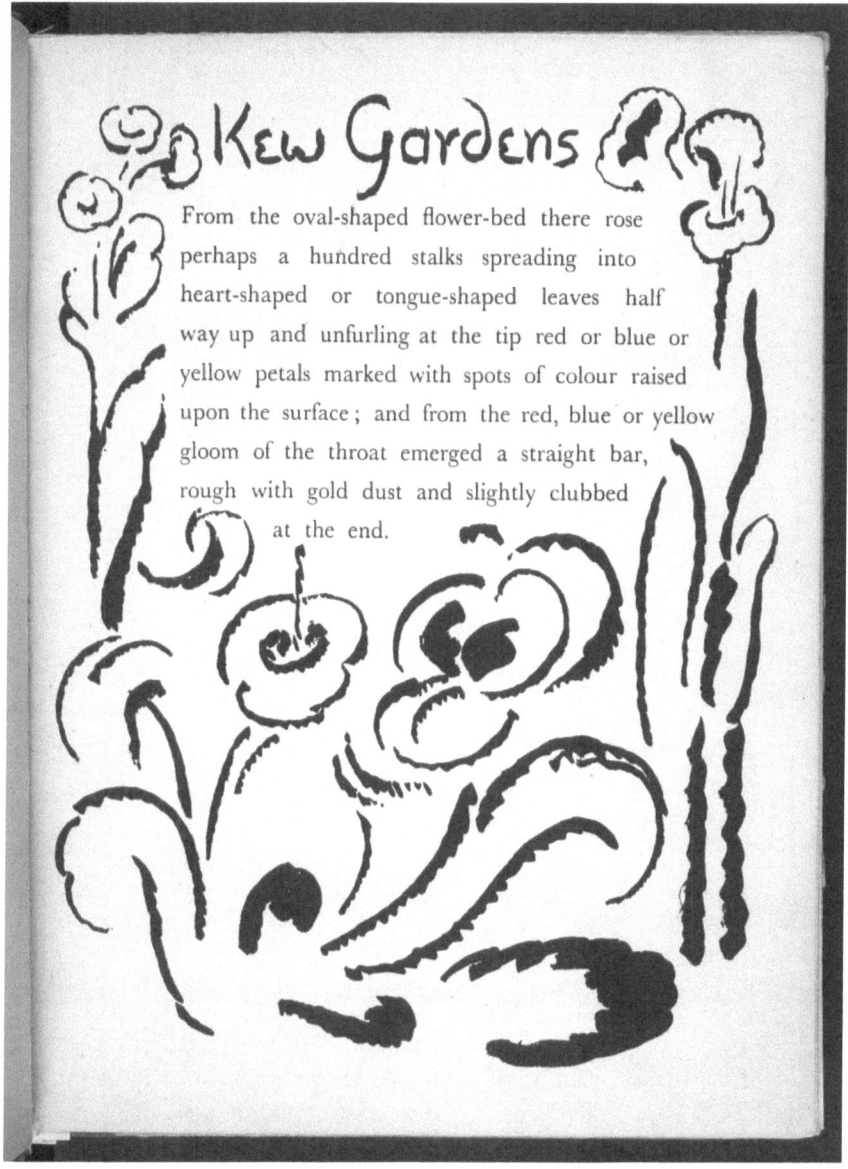

Figure 7.4 Vanessa Bell, cover for Virginia Woolf, Kew Gardens, 1927, p. 4. Photo courtesy of the British Library. © 2023 Artists Rights Society (ARS), New York / DACS, London.

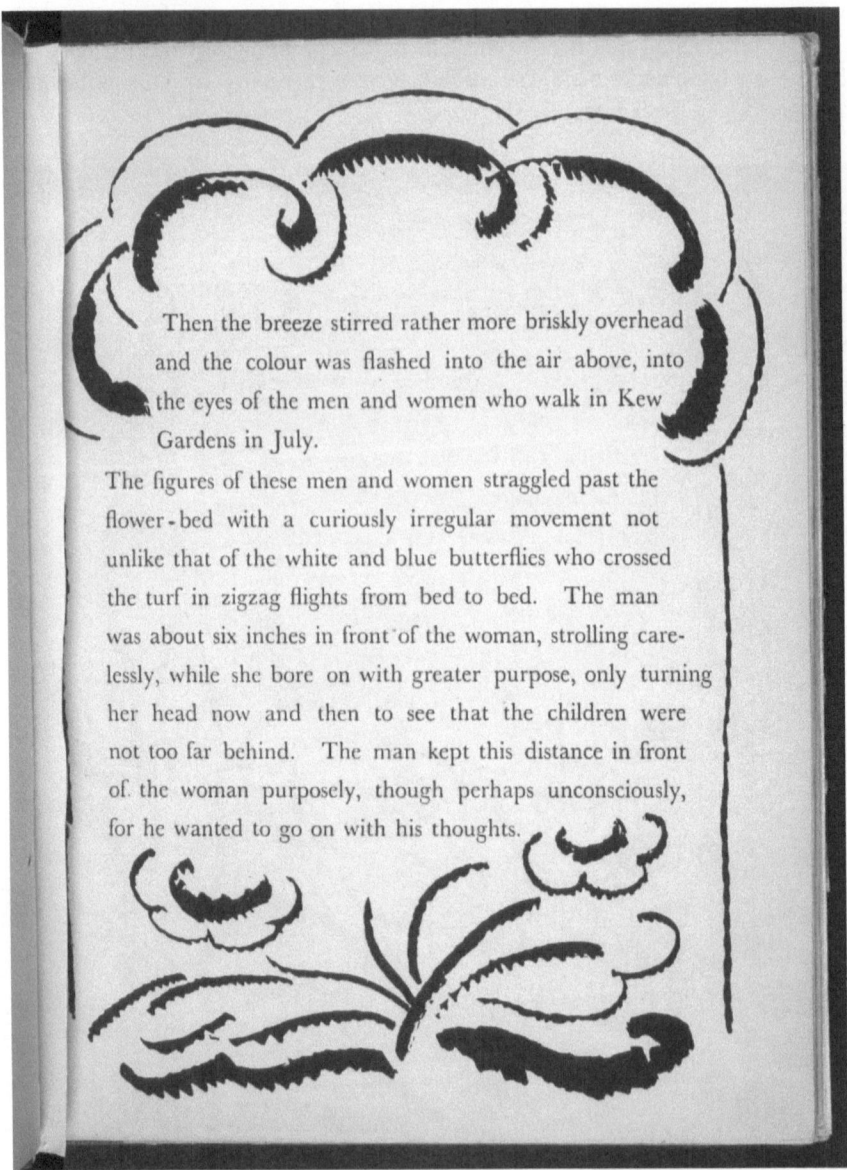

Then the breeze stirred rather more briskly overhead
and the colour was flashed into the air above, into
the eyes of the men and women who walk in Kew
Gardens in July.

The figures of these men and women straggled past the
flower-bed with a curiously irregular movement not
unlike that of the white and blue butterflies who crossed
the turf in zigzag flights from bed to bed.   The man
was about six inches in front of the woman, strolling carelessly, while she bore on with greater purpose, only turning
her head now and then to see that the children were
not too far behind.   The man kept this distance in front
of the woman purposely, though perhaps unconsciously,
for he wanted to go on with his thoughts.

Figure 7.5 Vanessa Bell, woodcut illustration in Virginia Woolf, Kew Gardens, 1927, p. 6.
Photo courtesy of the British Library. © 2023 Artists Rights Society (ARS), New
York / DACS, London.

We find an even more innovative interaction between text and image in the
woodcut matched with the paragraph about the third pair of visitors to the garden (Figure 7.6): the women who observe the eccentric elderly man while their
own words, unable to cohere into any discernible meaning, fall past the upright

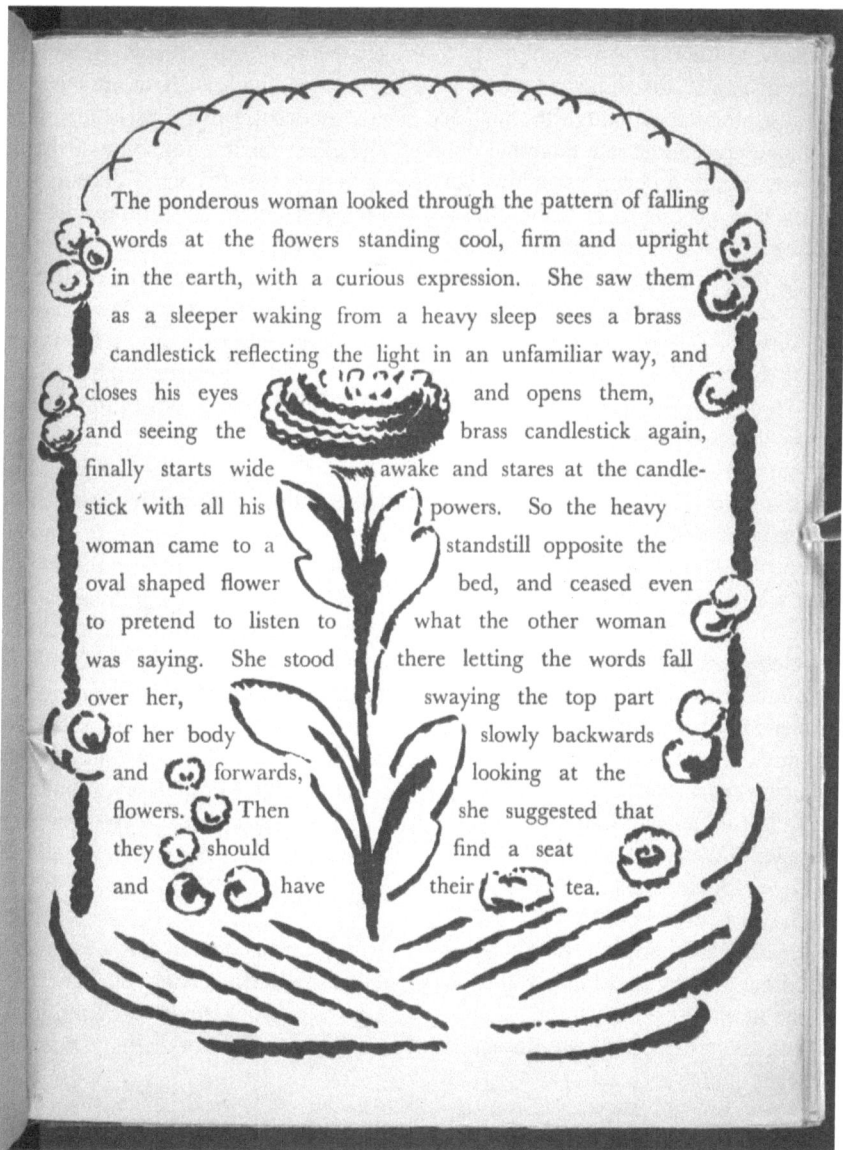

The ponderous woman looked through the pattern of falling words at the flowers standing cool, firm and upright in the earth, with a curious expression. She saw them as a sleeper waking from a heavy sleep sees a brass candlestick reflecting the light in an unfamiliar way, and closes his eyes and opens them, and seeing the brass candlestick again, finally starts wide awake and stares at the candlestick with all his powers. So the heavy woman came to a standstill opposite the oval shaped flower bed, and ceased even to pretend to listen to what the other woman was saying. She stood there letting the words fall over her, swaying the top part of her body slowly backwards and forwards, looking at the flowers. Then she suggested that they should find a seat and have their tea.

Figure 7.6 Vanessa Bell, woodcut illustration in Virginia Woolf, Kew Gardens, 1927, p. 16. Photo courtesy of the British Library. © 2023 Artists Rights Society (ARS), New York/DACS, London.

flowers. As in all these images for the second edition, Bell avoids any reference to the human figures—nor do we see the snail, walking stick, or thrush[74]—but here the "cool, firm and upright" flower takes on the nature of a responsible character, stalwart in the midst of the illusory "pattern of falling words" around it, suggested in the decorative circles descending to the earth below. Are they smaller blossoms, or seeds, hinting at the possibility of future growth out of the furrows below (out of the disparate, disconnected words might some meaning arise?), or are they barren abstractions? Although the imagery in this second set of prints is seemingly spontaneous, almost like marginal doodles, she experimented for some of the designs with half a dozen variations before reaching her final image.[75] Particularly noteworthy on this page is the effect of the insertion of Bell's central flower on the reader's ability to comprehend the flow of language. It first breaks the line of text in the phrase "closes his eyes | and opens them," setting up a rhythmic splitting that continues with the swaying of the woman's body, "slowly backwards | / and forwards," and concludes, finally, with that most mundane and comforting of activities, "and have | their tea" (though even that comfort is disrupted by the falling words/blossoms/seeds that further scatter the individual words of the text and their collective meaning).

Many of the woodcuts in the second edition are in fact purely abstract, but some, especially toward the end, seem like theatrical curtains or decorative proscenium arches, as if we are being invited into an alternative space of possibilities. The last woodcut (Figure 7.7) serves as a definitive conclusion, as we end with the angular lines and rigorously centralized mechanical wheel at the bottom; but industrialized London is joined with the delicate flowers in an arch above, with a suggestion of stippled grass and/or clouds. The top part drawn from nature is more old-fashioned than modernist in style, and this fusion of organic and mechanical forms, of nature and city, of nostalgia and modernism in Bell's last woodcut is illuminated by Tammy Clewell's argument that "modernist nostalgia entails an uncertain contestation between a 'perfect' past that is represented as an object of memory or phantasy and a 'tense' present figured in relation to the anxiety-producing developments of the age."[76] We are also reminded of Woolf's description of Lily Briscoe's painting in *To the Lighthouse* (1927):

> Beautiful and bright it should be on the surface, feathery and evanescent, one colour melting into another like the colours on a butterfly's wing; but beneath the fabric must be clamped together with bolts of iron. It was to be a thing you could ruffle with your breath; and a thing you could not dislodge with a team of horses.[77]

This takes us back to Roger Fry's ideas about the significance of an underlying form to provide structure and strength to a composition, evident in Lily's painting as well as in the text and images of *Kew Gardens*.

Woolf and Bell also overlap in certain ways with Fry in their theories about the role of illustrations, although he begins his essay "Book Illustration and a Modern Example" (1926) with a remarkable reference to warfare between the author and artist in the process of creating an illustrated book: "Book illustration is a battle ground, a no-man's land raked by alternate fires from the artist and the writer, claimed by both, sometimes nearly conquered by one, but only to be half recaptured by the other."[78] He writes that any kind of literal transcription of the author's words or ideas must be resisted, but he adds that "it may be possible to embroider

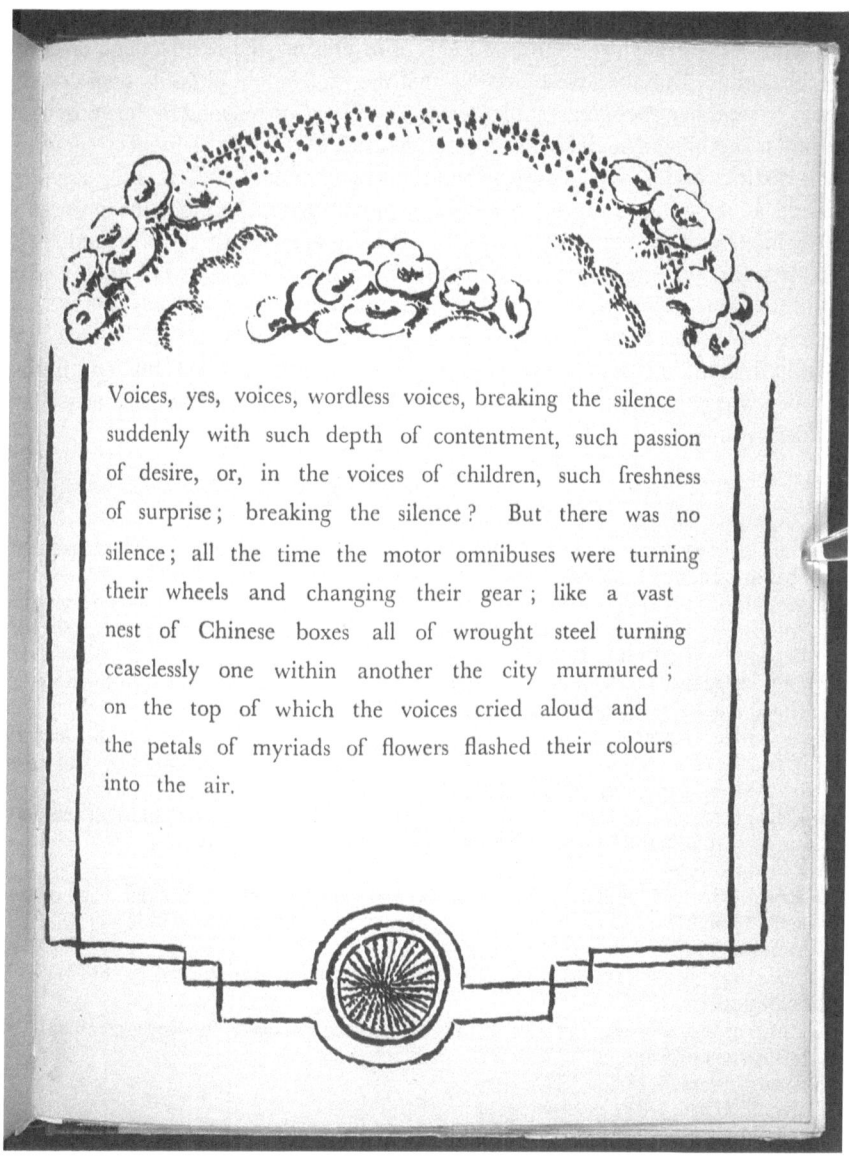

Voices, yes, voices, wordless voices, breaking the silence suddenly with such depth of contentment, such passion of desire, or, in the voices of children, such freshness of surprise; breaking the silence? But there was no silence; all the time the motor omnibuses were turning their wheels and changing their gear; like a vast nest of Chinese boxes all of wrought steel turning ceaselessly one within another the city murmured; on the top of which the voices cried aloud and the petals of myriads of flowers flashed their colours into the air.

Figure 7.7 Vanessa Bell, woodcut illustration in Virginia Woolf, Kew Gardens, 1927, p. 24. Photo courtesy of the British Library. © 2023 Artists Rights Society (ARS), New York / DACS, London.

the author's ideas or rather to execute variations on the author's theme . . .." The illustrator must be "discreet," however: "He merely starts a vague train of thought by the image which he puts before you in one of those pauses which the author's discursiveness allows."[79] Fry proposes that illustrations must be of the appropriate density, texture, linear freedom, and decorative style "to set the imagination free."[80]

As we have seen, both sisters agreed that the purpose of an illustration was not, in fact, to "illustrate"—i.e., to represent, or re-present, the word-image provided by the author. They believed, instead, that the image should be its own creative entity.[81] Together, then, text and image in *Kew Gardens* respond to the garden and its larger spatial and social context not as a series of discrete elements—plantings with distinct styles and histories, buildings with specific references, etc.—but instead as an opportunity to express a personal vision, explore the process of perception, and provide oblique social commentary. Woolf's words and Bell's woodcuts in both editions (though in quite different ways) allow our imaginations multiple entry points to the conflicted relationship among humans, nature, and the city, and their remarkable collaboration enabled both sisters to balance their own individual creative decisions with a more rigorous professionalism, finding a new sense of authority and freedom as they filled multiple roles as artists and businesswomen.

Notes

1 Both editions were published as short unpaginated books. For digital scans of the British Library copies, see https://www.bl.uk/collection-items/kew-gardens-by-virginia-woolf-1919 and https://www.bl.uk/collection-items/kew-gardens-by-virginia-woolf-1927. The story was also included in 1921 in her collection *Monday or Tuesday,* but without Bell's illustrations.

2 Qtd. in Jacques and Woudstra, *Landscape Modernism Renounced*, 92, from an unpublished manifesto co-authored by Jean Cannell-Claes. Tunnard later turned away from modernist landscape design and became a proponent of preservation, as Jacques and Woudstra discuss. Modernist trends in landscape design can be found somewhat earlier on the Continent; see, e.g., Haney, "Leberecht Migge."

3 Quaintance, "Kew Gardens," 33, referring to William Hogarth's satirical image of George III with the Pagoda at Kew in the background.

4 Hooker, "Report," 53.

5 Bruce, "Crown Colonies," 231. For representative pages on the imperial reach of Kew in two important recent studies, see Drayton, *Nature's Government*, 211, and Endersby, *Imperial Nature*, 135.

6 See Alexander, "Kew Gardens," 117, who writes that Kew "cannot be thought of as apolitical."

7 See Saguaro, *Garden Plots*, 12–13; also 14, where she describes Kew "as an axis and instrument of Empire."

8 Woolf, *Diary*, 1:28 and 127.

9 Woolf, *Diary*, 1:81 (23 Nov. 1917).

10 Woolf, *Orlando* 215.

11 Hancock, "Virginia Woolf and Gardens," 256. Woolf's descriptions of Kew in *Night and Day*, 329, provide a broader view of the gardens, with "the lake, the broad green space, the vista of trees with the ruffled gold of the Thames in the distance and the Ducal castle standing in its meadows."

12 Woolf, *Diary*, 1: 286 (3 July 1919). See also Hancock, "Virginia Woolf and Gardens," 256–57.

13 Susan Groag Bell, "Vanessa's Garden," 109–10 describes Bell as "not a driven gardener, but rather a painter who loved flowers and colors" (111; also 115). Still, Bell and Grant were "enthusiastic and practical gardeners"; Bell and Nicholson, *Charleston*, 128. See also Scott, *In the Hollow of the Wave*, 85–92 and Hancock, "Virginia Woolf and Gardens," 252–54.

14 Woolf, *Diary*, 1:55 (3 Oct. 1917). See Zoob, *Virginia Woolf's Garden*, on Monk House; also Scott, *In the Hollow of the Wave*, 92–99.

15 Letter from Woolf to Janet Case, *Letters*, 3:202. See also a letter from Sackville-West, 23 Apr. 1939: "I would so like Leonard to see my garden. You, I know, are no gardener, so I confine this interest to Leonard"; Sackville-West, *Letters*, 423.

16 Woolf, *Diary*, 1:40 (5 and 6 Aug. 1917). See also Alt, *Virginia Woolf*, 148–49.

17 Alt, *Virginia Woolf*, 147.

18 See Page and Smith, *Women, Literature, and the Arts*, for more discussion of eight British women, c. 1890–1940, who use the garden for both personal and professional growth. This essay is an extended variation of the epilogue to that book.

19 See Woolf, *Letters*, 2:364–66 and 369, and *Diary*, 1:280 (10 June 1919), for her delight at the sudden popularity of the first edition after this review; also Willis, *Leonard and Virginia Woolf*, 12, 32–33.

20 [Child], *Kew Gardens*, 293.

21 Forster, "Vision," 2; see also Staveley, "Conversations at Kew," 60, note 9.

22 Oakland, "Virginia Woolf's *Kew Gardens*," 264–68.

23 Staveley, "Conversations at Kew," 40, and Goldman, "Virginia Woolf," 41.

24 Mackay, "Fire and Broken Petals," n.p. See also Elliott and Wallace, *Women Artists and Writers*, 61 and 89, and Goldman, *Feminist Aesthetics* 150.

25 Staveley, "Conversations at Kew," 40 (who also gives an overview of some of the key literature on this story); Goldman, "Virginia Woolf," 41, 39; and Saguaro, *Garden Plots*, 7, 11.

26 [Child], *Kew Gardens*, 293.

27 Sorensen, *Modernist Experiments*, 214 and note 76 for her emphasis on the materiality of the book. The design of the 1919 cover was important to Woolf, who was interested in all details of bookmaking; she solicited ideas from Bell and Duncan Grant before deciding on Fry's hand-painted covers; *ibid.*, note 90.

28 Letter from Woolf to Bell (25 June 1918), *Letters*, 2:255, and letter from Bell to Woolf (3 July 1918), Vanessa Bell, *Selected Letters*, 214.

29 Letter from Woolf to Bell (8 July 1918), *Letters*, 2:258. Woolf later asks whether Bell could do the cover ("anything of course that you like, without reference to the story"), and adds, "your designs will be a tremendous draw." Letter to Bell (26 Nov. 1918), *Letters*, 2:298.

30 Letter from Woolf to Margaret Llewelyn Davies (2 May 1917), *Letters*, 2:151.

31 Letter from Woolf to Carrington (13 July 1917), *Letters*, 2:162–63. Despite Woolf's pleasure with Carrington's illustrations, the 1919 Hogarth Press edition lacked the woodcuts because of some technical issues. Woolf wrote to her on 18 June 1919, "I'm very sorry, as they added greatly to the charm of the work which will look very blank without them." *Letters*, 2:368.

32 The green insect is perhaps the Great Green Bush Cricket (*Tettigonia viridissima*).

33 Oakland, "Virginia Woolf's *Kew Gardens*, 268. Richter, *Virginia Woolf*, 85, in contrast, sees the snail as Woolf's "symbol of the 'victim'," related to "the story's theme of life as a phenomenon of exquisite but meaningless pattern and colour." See also Alexander, "Kew Gardens," 122 for an ecocritical perspective.

34 See Staveley, "Conversations at Kew," 55 and her note 11 for a range of critical responses to the impact of the Great War on this story.

35 Letters to Vanessa Bell (11 and 19 Nov. 1918), *Letters*, 2:290, 298. Woolf often neglected to include apostrophes, as in this sentence, and I follow the practice of the editors of her letters to omit putting [sic] after each example.

36 See Staveley, "Conversations at Kew," 50–52 for a convincing reading of gender and class issues here.

37 Alt, *Virginia Woolf*, 148 suggests that the image of the nesting boxes not only alludes to the interconnecting systems of the metropolis but also "refers backwards as well, suggesting the innumerable overlapping lives being lived in the gardens at Kew." See also Oakland, "Virginia Woolf's *Kew Gardens*," 273.

38 Saguaro, *Garden Plots*, 19, and Fry, "Essay in Aesthetics," 22.

39 Letter to Bell (1 July 1918), *Letters*, 2:257, and Forster, "Visions," 2. See also Stewart, "Impressionism," especially 41–43.

40 Perry, "Reminiscences," 120.

41 More specifically (and simplistically), the story progresses as follows: introductory paragraph (nature), first pair, snail, second and third pairs, snail, fourth pair, conclusion (city). Oakland, "Virginia Woolf's *Kew Gardens*," 266 notes Woolf's interest in Post-Impressionism and in Roger Fry's exhibition, and observes that "The various stages of the story appear to be very consciously planned in a formal and thematic attempt to create order despite (or because of) the fluid nature of the initial impressions."

42 Woolf, *Roger Fry*, 177–78. See also her letter to Violet Dickinson (27 Nov. 1910), *Letters*, 1:440. For the relationship between Bell and Fry, see Caws, *Women of Bloomsbury*, 71–91. The Second Post-Impressionist Exhibition (partly organized by Leonard Woolf) included 242 art works. In his short essay Clive Bell emphasizes "simplification and plastic design" as key elements in English Post-Impressionism, along with "significant form"; "English Group," 22.

43 Woolf, *Moments of Being*, 71–73. This essay was written 1939–40 but first published posthumously in 1972.

44 See Uhlmann, "Virginia Woolf," 58.

45 Fry, "Art and Science," 54. Uhlmann, "Virginia Woolf," 65–66 indicates the difference between the Impressionists' view of "sensation" and Cezanne's, where the eye is brought into conversation with the "organising mind."

46 For the impact of Post-Impressionist color on Woolf's writing, see Goldman, *Feminist Aesthetics*, 112–16.

47 Letter to Vita Sackville-West (16 March 1926), *Letters*, 3:247. Woolf also discusses Fry's emphasis on rhythm in her biography; *Roger Fry*, 214. See also Uhlmann, "Virginia Woolf," 61.

48 Gillespie, *Sisters' Arts*, 1–2 and Goldman, *Feminist Aesthetics*, 115–16. For more on the relationship between the two sisters, see Caws, *Women of Bloomsbury*, 48–59.

49 Woolf, *Recent Painting*, 2–3. Woolf particularly admired her sister's use of color, writing to Bell 4 Feb. 1940, "what a poet you are in colour . . ."; *Letters*, 6:381. See also Goldman, *Feminist Aesthetics*, 150, for the shared importance of non-verbal form and color for both women; also Susan Groag Bell, "Vanessa's Garden," 109.

50 Vanessa Bell, *Selected Letters*, 214. For an illustration of the painting, see the museum website: www.artandarchitecture.org.uk/images/gallery/f491cb65.html.

51 Letter to Bell, (12 May [1928]), *Letters*, 3:498.

52 Letter to Bell (15 July 1918), *Letters*, 2:259.

53 Letter to Bell (7 Nov. 1918), *Letters*, 2:289.

54 Gillespie, *Sisters' Arts*, 119 describes this as a "visual pun."

55 Woolf, *Mr. Bennett and Mrs. Brown*, 4. Cezanne was particularly important for Bell; Caws, *Women of Bloomsbury*, 107–10.

56 Gillespie, *Sisters' Arts*, 108–9, 268. See Robins, "'Manet and the Post-Impressionists'," for the checklist of works in this exhibition.

57 See Sorensen, *Modernist Experiments*, 218–20, where she emphasizes "the material text's play with scale," and her Fig. 5.9.

58 Woolf, *Diary*, 1:216 (9 Nov. 1918), and letter to Bell (6 June 1919), *Letters*, 2:365–66.

59 Bell's share of the profit came to only £3; letter from Woolf to Bell (4 June 1919), *Letters*, 2:365; also pp. 289, 297, 296, 298–99, 303, 352.

60 Woolf, *Diary*, 1:271 (12 May 1919).

61 Woolf, *Diary*, 1:279 (9 June 1919), and her letter to Bell, [4 June 1919], *Letters*, 2:365.

62 See Gillespie, *Sisters' Arts*, 118–37 for a useful discussion of Bell's woodcuts for the two editions of *Kew Gardens*, along with illustrations, and 332–33, note 21, for a lengthy account of the lack of critical attention to either these woodcuts or her other illustrations for Woolf.

63 See Bell and Nicholson, *Charleston*, for Bell's home at Charleston, and for the Bell's association with the Omega Workshops, see Spalding, *Vanessa Bell*, 122–29; also Lee, *Virginia Woolf*, 369–70.

64 Spalding, *Vanessa Bell*, 221.

65 Sorensen, *Modernist Experiments*, 220. The 1927 edition was printed by Herbert Reiach, London, who apparently did his typesetting within the frames of Bell's new set of woodcuts; Gillespie, *Sisters' Arts*, 125 and Willis, *Leonard and Virginia Woolf*, 389.

66 Qtd. in Sorensen, *Modernist Experiments*, 221.

67 Lee, *Virginia Woolf*, 369, and Harvey, "Lightness Visible," 111, 113.

68 Letter to Bell (15 July 1918), *Letters*, 2:259.

69 Woolf, *Walter Sickert*, 22. For more on Woolf's views about art, see Torgovnick, *Visual Arts*, 125–31.

70 Letter to Bell (17 Aug. 1937), *Letters*, 6:158; see also Bradshaw, "Virginia Woolf," 294 and Goldman, *Feminist Aesthetics*, 150.

71 Gillespie, *Sisters' Arts*, 125.

72 See Sorensen, *Modernist Experiments*, 221 for a somewhat different interpretation (with a window and urn or flowerpot being the stable elements).

73 Sorensen, *Modernist Experiments*, 223–24 provides a detailed analysis of this image; discussion of Bell's other prints and their relation to the text can be found in the following pages, 225–51.

74 One of Carrington's woodcuts for the Woolfs' *Two Stories* (1917) was a close view of a snail, which might explain why Bell decided not to illustrate that character in *Kew Gardens*.

75 Spalding, *Vanessa Bell*, 221.

76 Clewell, "Introduction," 3. See also Oakland, "Virginia Woolf's *Kew Gardens*," 267.

77 Woolf, *To the Lighthouse*, 194. Bell wrote to her sister, "surely Lily Briscoe must have been rather a good painter"; Vanessa Bell, *Selected Letters*, 318.

78 Fry, "Book Illustration," 157.

79 Fry, "Book Illustration," 158–59.

80 Fry, "Book Illustration," 166. He uses as his prime example E. McKnight Kauffer's images for *Burton's Anatomy of Melancholy* (1926), contrasting them with André Derain's woodcuts for Apollinaire's *L'Enchanteur Pourrissant*, which dominate the text. Thus, even though Derain's style "stimulates the imagination," there is no real partnership between text and image (172).

81 Gillespie, *Sisters' Arts*, 118 argues that despite Bell's willingness to subordinate her designs to her sister's text, she is able to "convey the fundamental psychological oppositions Woolf, too, tries to harmonize in her writing."

References

Alexander, Vera. "Kew Gardens as a Literary Space." *Studies in the History of Gardens & Designed Landscapes* 32, no. 2 (2012): 116–27.

Alt, Christina. *Virginia Woolf and the Study of Nature*. Cambridge: Cambridge University Press, 2010.

Bell, Clive. "The English Group." *Second Post-Impressionist Exhibition: British, French and Russian Artists*. London: Grafton Galleries, 1912.

Bell, Susan Groag. "Vanessa's Garden." In *Singular Continuities: Tradition, Nostalgia, and Identity in Modern British Culture*, edited by George K. Behlmer and F.M. Leventhal, 103–22. Redwood City, CA: Stanford University Press, 2000.

Bell, Quentin and Virginia Nicholson. *Charleston: A Bloomsbury House and Garden*. London: Frances Lincoln, 2004.

Bell, Vanessa. *Selected Letters of Vanessa Bell*. Edited by Regina Marler. New York: Pantheon, 1993.

Bradshaw, Tony. "Virginia Woolf and Book Design." In *Edinburgh Companion to Virginia Woolf and the Arts*, edited by Maggie Humm, 280–97. Edinburgh: Edinburgh University Press, 2010.

Bruce, Sir Charles. "The Crown Colonies and Places." *Proceedings of the Royal Colonial Institute* 36 (1905): 210–64.

Caws, Mary Ann. *Women of Bloomsbury: Virginia, Vanessa and Carrington*. New York: Routledge, 1991.

[Child, Harold]. Review of *Kew Gardens,* by Virginia Woolf. *Times Literary Supplement* 906 (29 May 1919): 293.

Clewell, Tammy. "Introduction: Past 'Perfect' and Present 'Tense': The Abuses and Uses of Modernist Nostalgia." In *Modernism and Nostalgia: Bodies, Locations, Aesthetics*, edited by Tammy Clewell, 1–22. London: Palgrave Macmillan, 2013.

Drayton, Richard. *Nature's Government: Science, Imperial Britain, and the 'Improvement' of the World*. New Haven, CT and London: Yale University Press, 2000.

Elliott, Bridget and Jo-Ann Wallace. *Women Artists and Writers: Modernist (Im)Positionings*. New York: Routledge, 1994.

Endersby, Jim. *Imperial Nature: Joseph Hooker and the Practices of Victorian Science*. Chicago: University of Chicago Press, 2008.

Forster, E. M. "Visions." *Daily News*, 31 July 1919: 2.

Fry, Roger. "An Essay in Aesthetics." 1909. In *Vision and Design*, 11–27. London: Chatto & Windus, 1920.

Fry, Roger. "Art and Science." 1919. In *Vision and Design*, 52–56. London: Chatto & Windus, 1920.

Fry, Roger. "Book Illustration and a Modern Example." In *Transformations: Critical and Speculative Essays on Art*. London: Chatto & Windus, 1926.

Gillespie, Diane Filby. *The Sisters' Arts: The Writing and Painting of Virginia Woolf and Vanessa Bell*. Syracuse, NY: Syracuse University Press, 1988.

Goldman, Jane. *The Feminist Aesthetics of Virginia Woolf: Modernism, Post-Impressionism, and the Politics of the Visual*. Cambridge: Cambridge University Press, 2001.

Goldman, Jane. "Virginia Woolf and Modernist Aesthetics." In *Edinburgh Companion to Virginia Woolf and the Arts*, edited by Maggie Humm, 35–57. Edinburgh: Edinburgh University Press, 2010.

Hancock, Nuala. "Virginia Woolf and Gardens." In *Edinburgh Companion to Virginia Woolf and the Arts*, edited by Maggie Humm, 245–60. Edinburgh: Edinburgh University Press, 2010.

Haney, David Henderson. "Leberecht Migge (1881–1935) and the Modern Garden in Germany." Ph.D. dissertation. University of Pennsylvania, 2005. Scholarly Commons. https://repository.upenn.edu/dissertations/AAI3165690.

Harvey, Benjamin. "Lightness Visible: An Appreciation of Bloomsbury's Books and Blocks." In *A Room of One's Own: Bloomsbury Artists in American Collections*, edited by Nancy E. Green and Christopher Reed, 88–117. Ithaca, NY: Herbert F. Johnson Museum of Art, 2008.

Hooker, Sir W.J. "Report from Sir W. J. Hooker, on the Royal Botanic Gardens, and the Proposed New Palm House at Kew," 30 Dec. 1844. *Accounts and Papers of the House of Commons* 45 (1845): 49–56.

Jacques, David, and Jan Woudstra. *Landscape Modernism Renounced: The Career of Christopher Tunnard (1910–1979)*. New York: Routledge, 2012.

Lee, Hermione. *Virginia Woolf: A Biography*. London: Chatto & Windus, 1996.

Mackay, Saffron. "Fire and Broken Petals: How the Suffragettes Made Their Mark on Kew."

Royal Botanic Gardens Kew, 2 Feb. 2018. https://www.kew.org/read-and-watch/fire-and-broken-petals-how-the-suffragettes-made-their-mark-on-kew.

Oakland, John. "Virginia Woolf's *Kew Gardens*." *English Studies* 68, no. 3 (1987): 264–73.

Page, Judith W. and Elise L. Smith, *Women, Literature, and the Arts of the Countryside in Early Twentieth-Century England*. Cambridge: Cambridge University Press, 2020.

Perry, Lila Cabot. "Reminiscences of Claude Monet from 1889–1909." *The American Magazine of Art* 18, no. 3 (1927): 119–26.

Quaintance, Richard. "Kew Gardens 1731–1778: Can We Look at Both Sides Now?" *New Arcadian Journal* 51 (2001): 14–51.

Richter, Harvena. *Virginia Woolf: The Inward Voyage*. Princeton, NJ: Princeton University Press, 1970.

Robins, Anna Gruetzner. "'Manet and the Post-Impressionists': A Checklist of Exhibits." *The Burlington Magazine* 152 (Dec. 2010): 782–93.

Sackville-West, Vita. *The Letters of Vita Sackville-West to Virginia Woolf*, edited by Louise DeSalvo and Mitchell A. Leaska. New York: William Morrow, 1985.

Saguaro, Shelley. *Garden Plots: The Politics and Poetics of Gardens*. New York: Routledge, 2006.

Scott, Bonnie Kime. *In the Hollow of the Wave: Virginia Woolf and Modernist Uses of Nature*. Charlottesville: University of Virginia Press, 2012.

Sorensen, Jennifer Julia. *Modernist Experiments in Genre, Media, and Transatlantic Print Culture*. New York: Routledge, 2016.

Spalding, Frances. *Vanessa Bell*. Boston: Ticknor & Fields, 1983.

Staveley, Alice. "Conversations at Kew: Reading Woolf's Feminist Narratology." In *Trespassing Boundaries: Virginia Woolf's Short Fiction*, edited by Kathryn N. Benzel and Ruth Hoberman, 39–62. London: Palgrave Macmillan, 2004.

Stewart, Jack F. "Impressionism in the Early Novels of Virginia Woolf." *Journal of Modern Literature* 9, no. 2 (May 1982): 237–66.

Torgovnick, Marianna. *The Visual Arts, Pictorialism, and the Novel: James, Lawrence, and Woolf*. Princeton, NJ: Princeton University Press, 1985.

Uhlmann, Anthony. "Virginia Woolf and Bloomsbury Aesthetics." In *Edinburgh Companion to Virginia Woolf and the Arts*, edited by Maggie Humm, 58–73. Edinburgh: Edinburgh University Press, 2010.

Wickham, Louise. *Gardens in History: A Political Perspective*. Oxford: Windgather Press, 2012.

Willis, John H. *Leonard and Virginia Woolf as Publishers: The Hogarth Press, 1917–41*. Charlottesville: University of Virginia Press, 1992.

Woolf, Virginia. *Monday or Tuesday*. London: Hogarth Press, 1921.

Woolf, Virginia. *The Diary of Virginia Woolf*. Edited by Anne Olivier Bell. 2 vols. New York: Harcourt Brace Jovanovich, 1977.

Woolf, Virginia. Foreword to *Recent Paintings by Vanessa Bell*. London: The London Arts Association, 1930.

Woolf, Virginia. *Kew Gardens*. London: Hogarth Press, 1919.

Woolf, Virginia. *Kew Gardens*. London: Hogarth Press, 1927.

Woolf, Virginia. *Mr. Bennett and Mrs. Brown*. London: Hogarth Press, 1924.

Woolf, Virginia. *The Letters of Virginia Woolf*. Edited by Nigel Nicolson. 6 vols. London: Hogarth Press, 1976.

Woolf, Virginia. *Moments of Being*. 1972. Boston: Houghton Mifflin Harcourt, 1985.

Woolf, Virginia. *Night and Day*. 1919. New York: George H. Doran, 1920.

Woolf, Virginia. *Orlando: A Biography*. 1928. New York: Harcourt, 2006.

Woolf, Virginia. *Roger Fry: A Biography*. London: Hogarth Press, 1940.

Woolf, Virginia. *To the Lighthouse*. 1927. London: Penguin, 1974.

Woolf, Virginia. *Two Stories*. London: Hogarth Press, 1917.

Woolf, Virginia. *Walter Sickert: A Conversation*. London: Hogarth Press, 1934.

Zoob, Caroline. *Virginia Woolf's Garden: The Story of the Garden at Monk's House*. London: Jacqui Small, 2013.

Epilogue

What If We Start with the Garden?

Judith W. Page

In our concluding thoughts for this volume, we want to glance back to the recent tradition of writing about nature and forward to new ways in which the collaborative art of gardens might be brought to the larger and more urgent world of nature in 2023. We have seen that garden theorists anticipate many of the ideas expressed in wider studies of nature, ecology, and environment, and that gardens involve the feminist values of collaboration and intersectionality as well as various forms of literary and artistic representation. As the noted American gardener and landscape architect Elizabeth Lawrence affirmed in *The Little Bulbs* (1957), "No one can garden alone."[1] Furthermore, people may also garden with the larger community in mind as an act of resistance in dark times, as a way of connecting with humanity through the beauty of the garden. In *Orwell's Roses* (2021), Rebecca Solnit reminds us that the author of *1984* gardened defiantly in the face of history.[2] And, indeed, our authors have introduced various forms of collaboration—gardening partners, garden clubs, community gardens—that reveal the interconnections that gardeners cherish. Because of the interplay of art and nature, gardens raise many questions about human agency and our responsibility to the natural world and to others, as well as our creative possibilities as humans. As Peter Remien and Scott Slovic posit in their introduction to *Nature and Literary Studies* (2022), "nature is an intersectional concept that draws together diverse material, discourses, and philosophical assumptions. Nature ... is central to how we understand agency, relationships, and humanity's place in the world."[3] Or, we might say, we cannot understand nature without human agency, without art.

As is often the case, some of the best clues about the future require us first to look back. Over thirty years ago, Michael Pollan published his influential book *Second Nature: A Gardener's Education* (1991) in which he argued for this dynamic interplay of art and nature: the garden "teaches me to know this patch of land more intimately, its geology and microclimate, the particular ecology of its local weeds and animals and insects."[4] In addition to teaching us (for Pollan includes "us" in the lesson) about the "particularities of place," Pollan says,

> Gardens also teach ... that there might be some middle ground between lawn and forest—between those who would complete the conquest of the planet in the name of progress, and those who believe it's time we abdicated our rule

DOI: 10.4324/9781003381549-9

and left the earth in the care of its more innocent species. The garden suggests there might be a place where we can meet halfway.[5]

For Pollan, then, as for other writers and theorists who follow him, gardens teach us because they can be sites of collaboration between the human and natural worlds. Following Pollan, David E. Cooper asks in *A Philosophy of Gardens* (2006),

> What if ... we were to take the relationship to the natural world that gardeners have as the starting point for reflection on our treatment of nature ... and work outward from there towards an appropriate stance toward wilder places?[6]

We hope to have demonstrated that gardens and gardeners can be the starting point, and that a range of feminist scholars, dedicated to collaboration, have taken up this "What if" question in their explorations of gardens, both real and imaginary.

Wallace Stevens meditates on this relationship between art and nature, artifact and wilderness, in "Anecdote of the Jar" (1919), a poem to which several garden theorists have alluded but not exhausted in their analysis.[7] The poem, condensed and enigmatic, consists of three quatrains:

I placed a jar in Tennessee,
And round it was, upon a hill.
It made the slovenly wilderness
Surround that hill.

The wilderness rose up to it,
And sprawled around, no longer wild.
The jar was round upon the ground
And tall and of a port in air.

It took dominion everywhere.
The jar was gray and bare.
It did not give of bird or bush,
Like nothing else in Tennessee.

A favorite among teaching anthologies, this poem always challenges students to ponder the complexities of language and metaphor, as well as the slipperiness of meaning. Rich with possibility, "Anecdote" provides a fertile ground for speculating on artifacts and nature. Stevens suggests a relationship in which art or artifice has no humility in the face of the wild, here represented by a mythical version of Tennessee. The jar is haughty—"of a port in air"—and it dominates the landscape. Rather than a harmonious creation, Stevens presents a cautionary tale about art's dominion. The plot of this anecdote is not a happy one of reconciliation or sharing. "Anecdote" offers a powerful warning: this explosive little poem presents a negative example in which both the artifact ("gray and bare") and the wilderness are diminished by the jar's presence rather than enhanced and enriched. Taking dominion everywhere is the opposite of an ecological sensibility or the "vocation of care" that Robert Pogue Harrison has identified as the hallmark of the gardener.[8] True gardeners enrich, giving back more than they take away. By extension, caring gardeners understand their power to transform nature and approach their work with humility rather than by dominion. Nature is not a wild, feminized land to be conquered, to borrow Annete Kolodny's foundational trope from *The Lay of the Land* (1975).[9]

Whereas David E. Cooper emphasizes the role of the "I" of Stevens' poem and the way in which "his" perceptions of nature change in the course of observation, John Dixon Hunt notes that "Gardens, too, are jars set down in otherwise untouched landscape, and part of their function and interest is that they alter their surrounding by their presence."[10] Yet "Anecdote" challenges us in more complex ways: it is not just about the speaker's changing perceptions; nor is Stevens' jar merely a symbol of the garden. Instead, the poem sets up a dynamic process of dominion rather than collaboration. If the jar represents a garden, then it is a failed place that does not account for or respect its surroundings—here culture does not collaborate with nature in order to meet halfway, in Pollan's terms. As with Vergil's *Georgics*, we are confronted yet again with a poem that misses the point of collaboration. A garden, of course, changes its surroundings, but its most humane theorists argue that it should not dominate and impose its will on those surroundings. As Wordsworth said, art must work hand in hand with nature; early on, he expressed a "gratitude toward insensate things" that distinguishes the true gardener and naturalist from Stevens' arrogant persona.[11] A reader steeped in the Romantics, Stevens would be aware of the dangers of overreaching in various contexts: if the "I" in "Anecdote" is the creative being, he might issue his decree to bring a world into being or attempt God-like dominion, but like Coleridge's Kubla Khan, he is more vulnerable than he thinks.

We have also seen in this volume that gardens and our perception of what gardens mean change over time—they have vital afterlives, to echo Hunt in his book *The Afterlife of Gardens* (2004). The afterlives of gardens exist in our experience of them and help transform them into various forms of art—they endure in artwork, films, novels, photographs, and poems—in the media that we have included in this volume. The arts thus collaborate with the natural record to create this afterlife. We have seen writers who leave records of their gardens, painters who imaginatively recreate the ruined gardens of Pompeii, French tapestries that tell the story of many lost interior decorations, filmmakers who negotiate their relationship to the natural world through their lens, and photographs that tell of the persistence and shared vibrancy of African American gardens in the rural South. The process of the transformation of actual gardens into art involves an active process of collaboration, sharing, and telling, all to create stories that endure.

Gardens are our starting point into the future of thinking about nature, as Kimberley Ruffin suggests in her exploration of the experience of African Americans in nature, *Black on Earth* (2010). Ruffin describes asking her students at Bates

College to reflect on the question of what is green and sustainable, on the meaning of ecology: "We decided that the urban, community garden, a space that melds rural and metropolitan realities, human and nonhuman nature, would be our muse. The garden seemed the perfect place to get beyond the human/nature divide that frustrated us all."[12] It is telling that Ruffin and her students find the garden the "perfect place" for this exploration, as if in response to Pollan and to Cooper's "What if" question. And it is also the place to think about the larger implications of our responsibility as humans in the world: "Ecological citizenship has the potential to foster ecological awareness and entitlement that moves people to protect not only their interest but also the interests of other humans and nonhumans to whom they are interconnected."[13] Ruffin's call for a shared commitment to humans and nature resonates with other forward-thinking writers, such as the late feminist writer bell hooks, who posits that our connection with nature is "humanizing" and believes in "the connectedness of all life."[14]

Other contemporary writers emphasize the role of race and Indigenous cultures in our understanding of nature, ensuring that the way we think about nature and gardens is more inclusive going forward. In *Black Faces, White Spaces*, Carolyn Finney prefers to use the term "environment" but she shares many of the values that Ruffin espouses in her writing on ecology. Finney notes that the "collective memory" of race informs attitudes toward the environment, and she values the "voicings" or personal stories that people tell of their experience in nature.[15] As we saw in Vaughn Sills' photographs of African American gardens, images, too, tell many such stories. Finney emphasizes that rather than ownership, which has often been denied to Black and Indigenous people, relationships to the natural world should be the standard.[16] This commitment to relationship also animates Jessica Hernandez's recent work with Indigenous environments in *Fresh Banana Leaves*, where she notes that planting banana trees provided life-giving food to resistance fighters in El Salvador and that her own father's healing from war began when he "began reclaiming his relationship with nature in the diaspora."[17] We are reminded of the solace that George Orwell found in his roses.

Although Hernandez does not refer explicitly to gardening, her discussion of conservation in Indigenous cultures brings us back to the garden, if indirectly. Hernandez explains that there is no word for "conservation" in Indigenous languages because before colonization the practice was not called for, but that "*taking care of* or *healing* are the closest words that come to mind when we try to explain conservation in our Indigenous languages."[18] We think here of Harrison's "vocation of care" as being essential to the gardener's ethos; this vocation is also related to what Carol Gilligan called a feminist "ethic of care," a relational way of thinking and living in the world.[19] The garden, so often dependent on women's collaborative efforts, can be a place of healing and a refuge from strife and at the same time a place of defiant resistance on behalf of the community. The garden is indeed our starting point for thinking about the natural world because the garden teaches us how to be in the world and how to let the world be.

Notes

1 Lawrence, *The Little Bulbs*, 5.
2 See also McKay, who writes broadly in *Radical Gardening* about the liberatory power of gardens for marginal groups, noting, for instance, the suffragists who destroyed the glass house at Kew (106) and queer activists in Britain who worked through the Pansy Project (147) or "guerilla" gardeners who planted a neglected patch of urban ground "*without permission*" (184). More recently, see Thomas, *The Intersectional Environmentalist*, for a "how-to" guide to environmental activism, especially 3–13.
3 Remien and Slovic, "Introduction: The Nature of Literature," 5.
4 Pollan, *Second Nature*, 75.
5 Pollan, *Second Nature*, 76–77.
6 Cooper, *A Philosophy of Gardens*, 107.
7 First published in the collection, *Harmonium*, 1919.
8 Harrison, *Gardens: An Essay on the Human Condition*, xi.
9 For a good overview on the history of feminist ecocriticism, see Gaard, "Nature, Gender, Sexuality," 261–79.
10 Cooper, *A Philosophy of Gardens*, 57, and Hunt, "The Garden as Cultural Object," 19.
11 Wordsworth, *Guide to the Lakes*, 74; *Letters of William and Dorothy Wordsworth: The Early Years*, 274–75
12 Ruffin, *Black on Earth*, 171.
13 Ruffin, *Black on Earth*, 167. Ecological citizenship is prefigured in Leopold's concept of the "land ethic," *Sand County Almanac*, 237–64.
14 hooks, "Earthbound," 186.
15 Finney, *Black Faces, White Spaces*, 10, 15.
16 Finney, *Black Faces, White Spaces*,115; see also Gundaker, "Introduction: Home Ground," 15.
17 Hernandez, *Fresh Banana Leaves*, 24–25, 28.
18 Hernandez, *Fresh Banana Leaves*, 76, emphasis in original.
19 Gilligan, *In a Different Voice*, 1982. Gilligan argues that "the logic underlying an ethic of care is a psychological logic of relationships, which contrasts with the formal logic of fairness that informs the justice approach" (73).

References

Cooper, David E. *A Philosophy of Gardens*. Oxford: Clarendon Press, 2006.
Finney, Carolyn. *Black Faces, White Spaces: Reimagining the Relationship of African Americans to the Great Outdoors*. Chapel Hill: The University of North Carolina Press, 2014.
Gaard, Greta. "Nature, Gender, Sexuality," 261–79. *Nature and Literary Studies*. New York: Cambridge University Press, 2022
Gilligan, Carol. *In a Different Voice: Psychological Theory and Women's Development*. Cambridge, MA: Harvard University Press, 1982.
Gundaker, Grey. "Introduction: Home Ground." *Keep Your Head to the Sky: Interpreting African American Home Ground*, 3–24. Ed. Grey Gundaker. Charlottesville: University of Virginia Press, 1998.
Harrison, Robert Pogue. *Gardens: An Essay on the Human Condition*. Chicago: University of Chicago Press, 2008.
Hernandez, Jessica. *Fresh Banana Leaves: Healing Indigenous Landscapes through Science*. Berkeley, CA: North Atlantic Books, 2022.

hooks, bell. "Earthbound: On Solid Ground." *Colors of Nature: Culture, Identity, and the Natural World*, 184–87. Eds. Alison H. Deming and Laret E. Savoy. Minneapolis, MN: Milkweed Editions, 2011.

Hunt, John Dixon. "The Garden as Cultural Object." *Denatured Visions: Landscape and Culture in the Twentieth Century*, 19–32. Eds. Stuart Wrede and William Howard Adams. New York: The Museum of Modern Art, 1991.

Hunt, John Dixon. *The Afterlife of Gardens*. Philadelphia: University of Pennsylvania Press, 2004.

Lawrence, Elizabeth. *The Little Bulbs: A Tale of Two Gardens*. Durham, NC: Duke University Press, 1986.

Leopold, Aldo. *A Sand County Almanac*. New York: Ballantine, 1966.

McKay, George. *Radical Gardening: Politics, Idealism, and Rebellion in the Garden*. London: Frances Lincoln, 2011.

Pollan, Michael. *Second Nature: A Gardener's Education*. New York: Dell Publishing, 1991.

Remien, Peter and Scott Slovic. "Introduction: The Nature of Literature," 1–28. *Nature and Literary Studies*. New York: Cambridge University Press, 2022.

Ruffin, Kimberly. *Black on Earth. African American Ecoliterary Traditions*. Athens: University of Georgia Press, 2010.

Solnit, Rebecca. *Orwell's Roses*. New York: Viking, 2021.

Stevens, Wallace. *The Collected Poems of Wallace Stevens*. New York: Alfred A. Knopf, 1975.

Thomas, Leah. *The Intersectional Environmentalist: How to Dismantle Systems of Oppression to Protect People + Planet*. New York: Voracious/Little Brown & Co., 2022.

Wordsworth, William. *Guide to the Lakes*. 5th ed. (1835). Ed. Ernest de Selincourt. New York: Oxford University Press, 1977.

Wordsworth, William and Dorothy Wordsworth. *The Letters of William and Dorothy Wordsworth: The Early Years, 1787–1805*. 2nd ed. Ed. Ernest de Selincourt. Rev. Chester L. Shaver. Oxford: Clarendon, 1967.

Index

Note: Numbers in *italics* refer to Figures.

Addams, J. 16
Addison, J. 104
African Americans 173–4; garden design
    as feminist ground 10–11, 13,
    14–18, 19–23, 24, 26; photographs
    of traditional African American
    gardens (Sills) (*see separate entry*);
    slavery 13, 15, 24, 113, 116
Allen, J.M. 129
Alma-Tadema, Lawrence 34–5, *35*, 37,
    38–9, 40, *40*, 41–2, 43–8, *44*, *45*,
    51, 52–5
Alt, C. 146, 147
Arbegast, M. 26
Arbenz, J. 141
archaeology and art *see* Pompeii, gardens
    of ancient
Arévalo, J.J. 141
Augusta, Princess 145
Austen, J.: *Pride and Prejudice* 101

Baker, E. 20
Barneby, R. 130, 131, 132
Bazzani, L. 34–5, 37, *38*, 39, *39*, 40, 41–2,
    43–4, 48–51, *48*, *49*, *50*, 52–5
'becoming' and 'becoming with' 10, 11,
    12, 14
Bell, A. 117, *120*
Bell, C. 151
Bell, V. *see Kew Gardens* (Virginia Woolf,
    with woodcuts by Vanessa Bell)
Benchoam, S. 136, 140
Bergmann, B. 105
Bethune, M.McL. 15
Blake, W. 157
Bliss, R. and M. 24
Blume, D. 75

Brakhage, S. 130–1; *The Wonder Ring*
    (1955) 130
Brooks, G. 15
Bryullov, K. 40
Buck, P. 17
Bullard, E. 22, 25
Bulwer-Lytton, E. 40, 53–4
Burnett, F.H. 12

Camille, M. 69
Capellanus, A. 69
care, ethic of 174
Carrington, D. 148
Catullus 47
Cezanne, P. 152, 154
Chard, C. 54
Charles V 66–8, 69, 72, 73–4, 75
Charles VI 69
Charles the Bold, Duke 78, 80
Chaucer, G. 104
Child, H. 147, 148
Christine de Pizan 69
civic improvement and transformational
    leadership 16–18
class 5, 11, 17, 18, 19, 20, 34, 39, 41,
    52, 113; Nashashibi, R.: *Vivian's
    Garden* (2017) 138, 139–40, 141;
    Valois France (circa 1365–1420)
    66, 69, 71
Clement V 74–5
Clewell, T. 162
clubs, garden 2, 18–21
Coffin, M.C. 25
collaboration, use of term 3
colonialism/imperialism 4, 43, 141, 174;
    Kew Gardens 145, 146, 147–8;
    post- 139

Comes, O. 42, 43, 45
Cooper, D.E. 172, 173, 174
Coste, J. 75
courtliness *see* Valois France (circa
        1365–1420)
Cran, M. 13
Crase, D. 132

Dale, P. 94
Daniels, L. 117, *119*
De Petra, G. 43
Dean, R. 12, 25
Deschamps, E. 69
design as feminist ground, garden 5, 10–26;
        'becoming' and 'becoming with'
        10, 11, 12, 14; civic improvement
        and transformational leadership 16–
        18; garden clubs 18–21; profession
        of landscape architecture 11, 12,
        16, 18–19, 21–6; tapestry that is the
        garden 12–16
Documenta 14 136–7
Dovzhenko, A. 130
DuBois, W.E.B. 14, 15

Ebers, G. 46
ecofeminist literary criticism 4
ecological citizenship 174
Edney, S. 2
Edwards, C. 54
Eisenhower, D.D. 141
Eisenstein, S. 130, 131
El Salvador 174
ethic of care 174
Evelyn, C. 12

feminism 4, 130, 131, 142, 147, 171, 172;
        design as feminist ground, garden
        (*see separate entry*); ethic of
        care 174
Ferrari, C.E. 10, 15
films: Menken, M.: *Glimpse of the Garden*
        (1957) (*see separate entry*);
        Nashashibi, R.: *Vivian's Garden*
        (2017) (*see separate entry*)
Finney, C. 174
Fiorelli, G. 35, 36, 37, 38, 42, 43, 55
Flanders, A.H. 16–18, *17*
Ford, R. 21
Forster, E.M. 147, 151
France 55; Valois France (circa 1365–1420)
        (*see separate entry*)
Franke, B. 77–8
Franklin, R. 3
Froissart, J. 69

Fry, R. 41, 148, 150, 151, 154, 157, 162–163
Fryar, P. *127*

Gaard, G. 4, 10
Gambart, E. 38
Gauguin, P. 154
Gautier, T. 53–4
Gell, W. 42
Georgics (Vergil) 1–2, 173
Gibson, R. 105
Giddings, P. 13
Gillespie, D.F. 152, 158
Gilligan, C. 174
Girard d'Orléans 75
Glave, D.D. 10, 13
Glickman, S. 113–14
Glimpse of the Garden see Menken, M.:
        Glimpse of the Garden (1957)
Godward, W. 52
Goldman, J. 147, 152
Goodrich, M.H. 16
Grant, D. 146
Gressin-Dumoulin de Boisgirard, M.-P. 45
Guatemala *see* Nashashibi, R.: *Vivian's
        Garden* (2017)
Gundaker, G. 116

Hamer, F.L. 20–1
Haraway, D.J. 11
Harris, D. 13
Harrison, R.P. 2, 97, 173, 174
Harvey, B. 158
Hawthorne, N. 97
Hazlitt, W. 100
Hernandez, J. 174
Hibberd, S. 106
Hidcote Manor 104
Holmes, M. 20
Hooker, W. 145
hooks, b. 10, 11, 13, 174
Horace 47
Hughes, L. 14
Hunt, J.D. 91, 99, 103, 173
Hurston, Z.N. 10, 14–15
Hutcheson, M.B.B. 18–19

Indigenous peoples/cultures 4–5, 44, 174;
        see also Nashashibi, R.: *Vivian's
        Garden* (2017)
intersectional concept: nature 171
Isabeau of Bavaria 76
Isabella of Portugal 80

Jackson, D.W. 116
Jean, Duke of Berry 66, 69, 72, 76, 77

Jean II the Good 75
Jensen, W. 54
John the Fearless, Duke 72–3
Johnson, J.W. 15
Johnson, W.R. 1
Johnston, L. 104
Jones, B. (later Farrand) 22, 23–4, *23*
Jones, J. 10, 13

Keats, J.: 'Ode to a Nightingale' 103
Kew Gardens (Virginia Woolf, with
        woodcuts by Vanessa Bell) 4, 7–8,
        145–64, *157*, *159*, *160*, *161*, *163*;
        atmospheric quality of 151–2; book
        reviews 147–8, 151; conflicted
        views of the garden 147; Empire
        145, 146, 147–8; ethology 147;
        experimental fiction 145, 147;
        gardens: extensions of interior
        spaces 146; Impressionist and
        Post-Impressionist art 150–1, 152;
        Orlando 146; political subtext
        147; printed commercially (1927
        edition) 156; printing press 148,
        156; 'rhythm' 152; scientific order/
        progress 145, 146; 'sensation' 152,
        154; story 148–50; structural clarity
        151–2, 162; *To the Lighthouse*
        162; woodcuts: extrapolations
        rather than illustrations 152, 164;
        woodcuts (1919 edition) 145,
        148, 152–6, *153*, *155*, 158, 164;
        woodcuts (1927 edition) 145,
        148, 152, 156–64, *157*, *159*, *160*,
        161, *163*
Kinzer, S. 141
Kolodny, A. 173

Landry, J. *125*
landscape architecture, profession of 11,
        12, 16, 18–19, 21–6
Lawrence, E. 2, 12, 171
Lawson, W. 12
Lee, H. 158
Lefalle-Collins, L. 116
Limbourg, Herman, Paul and Jean de: Très
        Riches Heures 66, *67*, 68, 69–72,
        70, *71*, 75, 77, 84
Livy 47
Logan, M.D. 12

Maas, W. 130–1
MacDonald, S. 130–1, 132–3
McKittrick, K. 10
McNear, S.A. 129

McWillie, J. 116
Malouel, J. 82
Margaret of Bavaria 81
Margaret of Flanders 76, 81–2, 84
Marx, L. 132–3
Matisse, H. 154
Max, A. 53
Mazois, F. 42
medieval literature and pictorial arts *see*
        Valois France (circa 1365–1420)
Mekas, J. 130
Menken, M.: *Glimpse of the Garden*
        (1957) 6–7, 129, 130–5, *134*,
        135, *136*, 142; birdsong: stock
        source 133; camera movement
        133–5; *Dwightiana* (1957) 131–2;
        feminism 130, 131, 142; filming
        location 131; juxtaposition of
        filmic shots in motion and montage
        131; magnifying lens 132;
        temporality 133
Monet, C. 151
Moore, E. 117, *126*
Morelli, D. 39
Morisot, B.: *Butterfly Hunt* 151
Morrison, T. 115
Mulvey, L. 138, 139, 140

Nash, S. 84
Nashashibi, R.: *Vivian's Garden* (2017)
        4, 6–7, 129–30, 135–42, *138*;
        caretaker, Don Tomás 138, 139,
        140; closed-off and sheltering
        aspects 139; commissioning of film
        136–7; elements of Guatemalan
        history 141; fear and rival drug
        gangs 140; feminism 130, 142;
        interior spaces and garden 138;
        menacing criminal neighbor 140;
        refuge and danger/healing and
        terror 140; shared mode of being
        142; temporality 139; working-
        class Guatemalan laborers 138,
        139–40, 141; worry about Vivian's
        upcoming trip to Greece 141–2
native plant movement 19, 25

Oakland, J. 147, 149
O'Keeffe, G. 135
Orwell, G. 171, 174
Ovid 47
Owen, E.M. 117, *122*

Peters, A. 20
Philip the Bold, Duke 81, 82, 83, 84

Philip the Good, Duke 76, 77, 78, 79, 80
Phillips, K. 26
photographs 129; of traditional African
    American gardens 4, 6, 112–19,
    119–27, 174
photographs of traditional African
    American gardens (Sills) 4, 6,
    112–19, *119–27*, 174; black-and-
    white 115; borders 116; circles 117;
    color blue 119; color white 117;
    composing 115; distinctiveness
    115–17; introduction 112–14; light
    113, 115, 117; memory and identity
    116; philosophy and religion of
    Yoruba and Kongo peoples 116,
    117; pipes 117; place for social
    gatherings 116; process of taking
    114–15; raking of earth 119;
    swept yard 117; transformative
    space 114; trust 112, 115, 117–18;
    upside-down or broken vessel 117;
    watchers 117; water 117
Piozzi, H. 100
Platt, C.A. 25
Pliny the Elder 51
Pliny the Younger 104–5
Pollan, M. 171–2, 173, 174
Pompeii, gardens of ancient 5, 34–55;
    Alma-Tadema's flowers *35*,
    44–8, *44, 45*; archaeological
    genre painting 37–41, *38, 39, 40*;
    Bazzani's courtyards 48–51, *48, 49,
    50*; birth of garden archaeology 43;
    culture of reconstruction 35–7, *36*;
    current beliefs regarding ancient
    gardens 53; female containment and
    domestication 54; female isolation
    52; feminised ancient past and ruins
    54; green spaces and anachronisms
    41–4; plotting Pompeian gardens
    52–5; slippage between accuracy
    and fantasy 55
Propertius 47
Pudovkin, V. 130

queer theory 131

race 2, 4, 173–4; garden design as feminist
    ground 10–11, 13, 14–18, 19–23,
    24, 26; photographs of traditional
    African American gardens (Sills)
    (*see separate entry*); slavery 13, 15,
    24, 113, 116
Ragona, M. 131–2, 133
Rawle, M.C. 23

Rehmann, E. 12
Remien, P. 171
Ripley, D. 130, 131–2
Robinson, B. 113, 117, *121*
Robinson, W. 107
Roncicchi, N. 43
Ruffin, K. 173–4
Ruggiero, M. 43
Ruskin, J. 41

Sackville-West, V. 104, 107, 146
Saguaro, S. 147, 150
Sargent, C.S. 23
Sauval, H. 73–4, 75, 76
Schlesinger, S. 141
schools 54
Scott, B.K. 146
Scott, R. 53
Shelton, L. 12
Shipman, E.B. 23, 24–5, *25*
Shub, E. 130
Sills, V. 15
Sissinghurst Castle Garden 104, 107
Slovic, S. 171
Sluter, C. 82–5
Sogliano, A. 43
Solnit, R. 171
Somervell, T. 2
Sommer, G. 36, 37, 40
Sophocles 3
Sorensen, J.J. 156
Soviet montage 130
Spalding, F. 158
Spano, G. 43
Spencer, A.B. 10, 14–16, *14*, 25
Spencer, E. 14–15, *14*
Staveley, A. 147
Stevens, W.: 'Anecdote of the Jar' 172–3
Sturghill, A.B. *123*
Suarez, J.A. 130
Suffragettes 147
suffragists 54, 147
Suter, V. 4, 7, 129–30, 135–42
Szymczyk, A. 137

Tabor, G. 12
Taussig, M. 140
Terrell, M.C. 19
Theophrastus 51
Thompson, H. *124*
Thompson, R.F. 116
Thoren, R. 18
Tibullus 47
transformational artifice of garden *see*
    Valois France (circa 1365–1420)

transformational leadership and civic improvement 16–18
trust 112, 115, 117–18
Tunnard, C. 145
Turin, V. 130
Tyler, P. 133

United Fruit Company 141
United Nations: Women's Fund for Gender Equality 139
United States 10, 11, 52; African Americans (*see separate entry*); America's Little House: garden design 16–18, *17*; *Better Homes in America Campaign* 17; garden clubs 18–21; Guatemala 141; landscape architecture 11, 12, 16, 18–19, 21–6; native plant movement 19, 25; slavery 13, 15, 24, 113, 116; village improvement movement (nineteenth century) 16

Valois France (circa 1365–1420) 5–6, 66–85; Ahasuerus and Esther 77–9, *78*; Carthusian foundation at Champmol 82–5, *83*; Chambre du Cerf (Stag Room) in Avignon 74–5, *74*, 76; château de Germolles in Burgundy *80*, 81–2, 84; cultured courtier and natural peasant 69; diffusion inside and outside 68; faux book 72; Hesdin in Artois 68; *hortus conclusus* (enclosed garden) 68, 80; hunt tapestries 75–6; Limbourg brothers: Très Riches Heures 66, *67*, 68, 69–72, *70*, *71*, 75, 77, 84; liminality 68; ludic space 68; peasants and natural world 72; poetry 69; ribaldry 69, 80; simulation or semblance of gardens: painted or tapestry interiors 68; Sluter 82–5; social code 66; *Songe du vergier* (Dream of the Orchard) 66–8, 77; *Speculum humanae salvationis* (Mirror of Human Salvation) 77; spiral staircase: tower of Hôtel d'Artois 72–3, *73*, 82; Turin-Milan Hours 76–7, *79*; *verdure* (roughly, greeneries) 79, 81
Van Eyck, J. 76
Van Gogh, V. 154
Van Rensselaer, M.G. 22

Vergil: *Georgics* 1–2, 173
Vertov, D. 130
village improvement movement (nineteenth century) 16
Vivian's Garden see Nashashibi, R.: *Vivian's Garden* (2017)

Walker, A. 10, 11, 12, 15, 116
Warner, A.B. 12–13
Washington, M.M. 19–20, 22
Waterhouse, J.W. 52
Welty, E. 112
Werve, C. de 82
Wharton, E. 23
Wild, E. 7, 129–30, 135–42
Williston, D. 22–3, 26
Woolf, L. 145, 146, 148, 151, 158
Woolf, V. 10, 12; *Kew Gardens* (Virginia Woolf, with woodcuts by Vanessa Bell) (*see separate entry*)
Wordsworth, William and Dorothy 5, 6, 91–107, 173; 'A Farewell' 107; 'A Flower Garden' 102–3; affective qualities of garden 95, 96, 106; afterlife of garden 93–6, 107; art works hand in hand with nature 100–1, 107; boundaries and enclosure 102–3, 105, 106; deep ecological connection 92; diminutive, power of the 92; garden artistry 96, 97; garden rooms 104–5; home-making 92–3; 'How Sweet It Is' 107; microcosm of natural world 100; 'Nuns Fret Not' 97, 103; pathways 103–4, 105; 'Poor Robin' 107; Preface to *Lyrical Ballads* 97, 100; smallness and bounded spaces 93, 97; statuary: Isola Bella in Lake Maggiore 97–100, *98*; temporality of gardens 93, 96, 102–3; 'The Massy Ways, Carried Across These Heights' 94–5, 104; *The Prelude* 92; 'The Ruined Cottage' 94; 'Tintern Abbey' 101; topography of the place 93; Town End garden (Dove Cottage) 91–3, 107; travel writing 98, 99–100; trimness and neatness 106–7; wild and cultivated nature 101–2; Winter Garden at Coleorton 93–4, 96, 100, 102–4

Ydígoras Fuentes, M. 141
Yen, B.C. 94

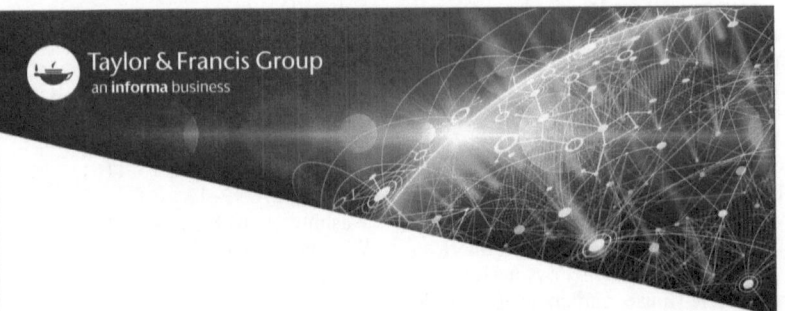
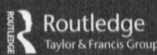
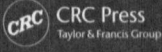